Mohamed Elzagheid

Water Chemistry, Analysis and Treatment

Also of Interest

Chemical Technicians.
Good Laboratory Practice and Laboratory Information Management Systems
Mohamed Elzagheid, 2023
ISBN 978-3-11-119110-2, e-ISBN 978-3-11-119149-2

Chemical Laboratory.
Safety and Techniques
Mohamed Elzagheid, 2022
ISBN 978-3-11-077911-0, e-ISBN 978-3-11-077912-7

Macromolecular Chemistry.
Natural & Synthetic Polymers
Mohamed Elzagheid, 2021
ISBN 978-3-11-076275-4, e-ISBN 978-3-11-076276-1

Bioremediation Technologies.
For Wastewater and Sustainable Circular Bioeconomy
Edited by Riti Thapar Kapoor, Mohd Rafatullah, 2023
ISBN 978-3-11-101665-8, e-ISBN 978-3-11-101682-5

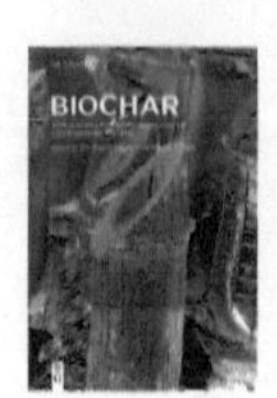

BioChar.
Applications for Bioremediation of Contaminated Systems
Edited by Riti Thapar Kapoor, Maulin P. Shah, 2022
ISBN 978-3-11-073858-2, e-ISBN 978-3-11-073400-3

Mohamed Elzagheid

Water Chemistry, Analysis and Treatment

Pollutants, Microbial Contaminants, Water and Wastewater Treatment

DE GRUYTER

Author
Prof. Dr. Mohamed Ibrahim Elzagheid
Chemical Engineering Department
Jubail Industrial College
Jubail Industrial City
Jubail 31961
Saudi Arabia
melzagheid@gmail.com;
elzagheid_m@rcjy.edu.sa

ISBN 978-3-11-133242-0
e-ISBN (PDF) 978-3-11-133246-8
e-ISBN (EPUB) 978-3-11-133261-1

Library of Congress Control Number: 2023945078

Bibliographic information published by the Deutsche Nationalbibliothek
The Deutsche Nationalbibliothek lists this publication in the Deutsche Nationalbibliografie; detailed bibliographic data are available on the Internet at http://dnb.dnb.de.

Cover image: Irina Vodneva/iStock/Getty Images Plus
Typesetting: Integra Software Services Pvt. Ltd.
Printing and binding: CPI books GmbH, Leck

www.degruyter.com

This book is dedicated to everyone who has worked day and night
to offer clean and safe water to people all across the world.

Preface

The book covers water chemistry, water sources, water pollutants, and microbiological contaminants. Water chemistry principles are explained in a straightforward manner. The book also looks at the theoretical foundations of several water treatment and analysis processes, as well as stormwater management and green infrastructure.

This book would be useful for graduate and advanced undergraduate students, as well as environmental researchers, chemists, and lab technicians working in water and environmental laboratories. Because the bulk of books on the market are aimed toward chemical engineers and operators, chemists and technicians alike can benefit greatly from this book. The book's language is clear, short, and easy to understand for all readers, including those with only a basic understanding of water chemistry.

I hope students, technicians, researchers, chemists, and scientists enjoy reading this book and find what they need to further their education.

Dr. Mohamed Ibrahim Elzagheid, Chemistry Professor
Waterloo, Ontario, Canada
2023

https://doi.org/10.1515/9783111332468-202

Acknowledgments

First and foremost, I want to express my gratitude to my entire family for always assisting and supporting me during my academic career.

I would also like to thank everyone in the Chemical Engineering Department at Jubail Industrial College, and special thanks to those who joined me in teaching the water and wastewater treatment course for regular students and Aramco company trainees in Saudi Arabia, as well as my former colleagues at the water desalination plant in north Benghazi, Libya.

Last but not least, a heartfelt thanks to the entire publishing team, particularly Ute Skambraks, Helene Chavaroche, and Suruthi Manogarane whose assistance and hard work cannot be overstated.

https://doi.org/10.1515/9783111332468-203

The Author

Mohamed Elzagheid is an Associate Professor of Chemistry at Jubail Industrial College (JIC), which is affiliated with the Saudi Royal Commission for Jubail and Yanbu. He also serves as a professor and consultant for the Libyan Authority for Scientific Research, which is associated with the Libyan Ministry of Education.

During his 23-year career at McGill University, SynPrep Inc. in Montreal (Canada), and JIC in Saudi Arabia, he was directly involved in the education of laboratory technicians and chemists, as well as supervised many undergraduate and graduate chemistry students, and has significantly contributed to numerous short-term and long-term training programs for chemistry-based laboratory technicians for local companies in Saudi Arabia.

He also served on JIC's Research, Publications, Projects, and Academic Promotion Team; Academic Promotion Committee; Curriculum Development Committee; Industrial Chemistry Technology Program Advisory and Evaluation Committee; CTAB Steering Accreditation Committee; Industrial Outreach Committee; and Chemical Engineering Department Safety Committee.

Dr. Elzagheid is the author of five textbooks that are now used to train chemistry-based technicians: *Introductory Organic Chemistry*, *Thoughts on Organic Chemistry*, *Macromolecular Chemistry: Natural and Synthetic Polymers*, *Chemical Laboratory Safety and Techniques*, and *Chemical Technicians: Good Laboratory Practice and Laboratory Information Management Systems*.

His work at Turku University in Finland, McGill University in Canada, and JIC in the Kingdom of Saudi Arabia has helped him establish a solid name in chemistry and chemical education in general, as evidenced by his research papers and publications.

https://doi.org/10.1515/9783111332468-204

Contents

Chapter 1 Introduction

1.1 Background

Water is a clear fluid that composes the world's streams, lakes, oceans, and rain as well as the primary component of organism fluids. A water molecule is a chemical substance composed of one oxygen atom and two hydrogen atoms linked by covalent bonds. At typical ambient temperature and pressure, water is a liquid, yet it coexists on Earth alongside its solid state, ice, and its gaseous state, steam.

Liquid water is essential for life on Earth because it acts as a solvent. It can dissolve molecules and enable important chemical reactions in animals, plants, and microbial cells. Because of its chemical and physical properties, it can dissolve more compounds than most other liquids. Every day, water is needed for a variety of purposes. We need water for a variety of reasons, including drinking, washing, cleaning, and cooking. There is no existence without water, and no life without air.

Water is locked up in the crystal lattices of minerals that make up rocks; it occurs in the innumerable microscopic pore spaces of rocks, from the surface to depths of over 5 km. Water is a simple enough substance, consisting of two hydrogens and one oxygen, yet it has some fascinating and essential features. Water's boiling and freezing temperatures are particularly high in comparison to other compounds of equal molecular weight, allowing it to exist at the Earth's surface in all three phases: solid, liquid, and vapor.

Although water is in a continuous cycle on Earth, it is consumed before it completes its cycle due to population expansion, environmental pollution, cost, mindless water usage, and climatic variations. It becomes increasingly challenging to obtain agricultural, industrial, drinking, and utility water.

Water, as a natural resource, is required for all living species to survive. The amount of water on the Earth's surface, however, is limited. The world's 1.4 billion km^3 of water is made up of 97.5% saline water and 2.5% fresh water. The percentage of freshwater distribution is 69.5% for polar glaciers, 30.1% for groundwater, and 0.4% for surface water resources.

In addition to issues with the number of freshwater supplies, there have also been issues concerning water quality on a worldwide scale. Rapid population expansion, technological advancements, urbanization, and global climate change are the primary causes of the deterioration of natural resources in terms of quantity and quality.

https://doi.org/10.1515/9783111332468-001

1.2 Water Importance

Water is essential for people and our prosperity. Because of its high dielectric constant, it is regarded as a universal solvent. Water accounts for about 60% of an adult's body weight on average ranging from 31% in bones to 83% in the lungs. Water performs numerous critical functions in the body, including waste removal, temperature regulation, nutrition transfer, and digestion. Water is the best hydrator for the body. It also moisturizes the tissues around the eyes, nose, and mouth. It safeguards the body's organs and tissues. It transports nutrients and oxygen to cells. Joint lubricant. Draining away waste materials reduces the stress on the kidneys and liver. Drinking water has many health benefits, some are illustrated in Figure 1.1.

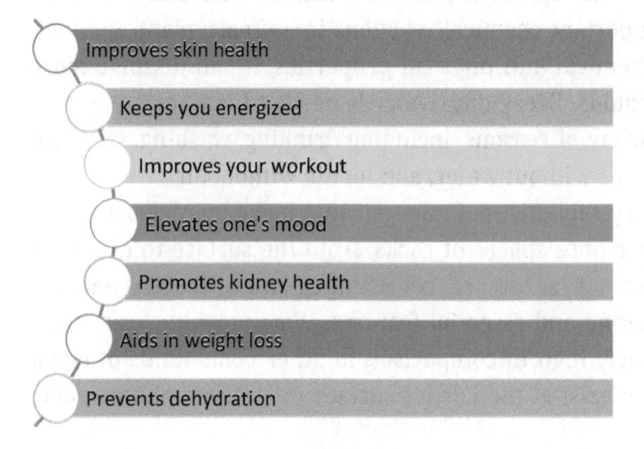

Figure 1.1: Drinking water health benefits.

Water is required by all living organisms including humans, plants, and animals to survive. It is actually necessary for life on the Earth to continue. Human bodies require water to help regulate body temperature and sustain all critical processes. Due to the body's loss of water through breathing, sweating, and digesting, it is vital to rehydrate and replenish water through beverages and foods containing water. Water is the most abundant component of saliva. It is required for the digestion of solid foods and the maintenance of oral health.

Maintaining a normal body temperature requires staying hydrated. When we sweat or are in a hot environment, our bodies lose water. Sweating keeps our bodies cool, but if we don't replenish the water we lose, our body temperatures rise. Dehydration results from a lack of water, which causes electrolyte and plasma levels to decline. Proper hydration is essential for maintaining optimal brain health. Inadequate water consumption has a negative impact on our focus, alertness, and short-term memory. Water helps to lubricate and cushion our joints, spinal cord, and tissues. This pushes us to be more physically active and relieves the pain produced by disorders such as arthritis. We all need water to restore fluids lost via sweating. Drinking enough water assists our kidneys to function more efficiently, lowering the risk of kidney stones.

It is critical to drink plenty of water while working out, participating in sports, or simply being active. Keeping hydrated has an impact on our strength, power, and endurance. Water assists to increase our metabolic rate. Exercising in the heat without staying hydrated can lead to major medical complications. In fact, severe dehydration can result in convulsions and, in some cases, death. Dehydration occurs when the body does not receive enough water. And, because water is essential for the proper functioning of many biological functions, dehydration can be quite harmful. It can even have fatal implications. Severe dehydration can have significant consequences. Drink plenty of water to avoid dehydration.

1.3 Safe Drinking Water Act

The WHO standards for safe drinking water employ a daily per capita consumption value of 2 L of drinking water for adults weighing 60 kg in the computation. A 10-kg child is considered to drink 1 L of water per day, while a 5 kg infant consumes 0.75 L/ day. The potable water that can be given to the user is suitable for drinking, food preparation, personal hygiene, and washing. At the point of supply to the consumers, this water must meet the requisite chemical, biological, and physical quality criteria. Any country's Safe Drinking Water Act is intended to ensure reliable and safe drinking water supplies while also protecting public health by regulating public water systems. This necessitates the formulation and enforcement of standards for public drinking water systems. To control water quality, we should consider controlling:
- microbiological caliber,
- chemical efficiency,
- by-products of disinfection and disinfectants,
- by-products of corrosion,
- chemicals used in agriculture,
- organic substances that are volatile, and
- radionuclides are radioactive unstable chemical elements that emit radiation as they degrade and become more stable. Radionuclides include radium-226, cesium-137, and strontium-90.

1.4 Water Contaminants or Pollutants

The term "contaminant" in the **S**afe **D**rinking **W**ater **A**ct (**SDWA**) is defined as any physical, chemical, biological, or radiological substance or matter in water. As a result, the legislation defines "contaminant" broadly as anything other than water molecules. Some pollutants could be reasonably expected to be present in drinking water. Some contaminants in drinking water may be dangerous if taken in specific quantities, while others may be innocuous. The presence of pollutants does not always imply that the water is unsafe to drink.

The **C**ontaminant **C**andidate **L**ist (**CCL**) contains just a subset of the universe of pollutants defined above. The CCL is the first stage of examination for unregulated drinking water contaminants that may require additional investigation of potential health impacts and levels observed in drinking water. The broad types of drinking water pollutants and examples of each are as follows:

- Physical contaminants have a great impact on the physical appearance and properties of water. Physical contaminants include sediment and organic detritus suspended in lakes, rivers, and streams.
- Substances or elements form chemical contaminants. These can be natural or man-made. Chemical contaminants include nitrogen, bleach, salts, pesticides, metals, bacterial toxins, and human or animal pharmaceuticals.
- Biological contaminants are organisms that live in water. Microbes and microbiological contaminants are other terms for them. Bacteria, viruses, protozoa, and parasites are examples of biological or microbiological contaminants.
- Radiological contaminants are chemical elements having an uneven number of protons and neutrons that produce unstable atoms capable of emitting ionizing radiation. Cesium, plutonium, and uranium are examples of radioactive contaminants.

1.4.1 Contaminant Candidate List

1.4.1.1 Microbial Contaminant Candidates

Bacteria, viruses, and fungi are examples of microbial contaminants. Some of the potential consequences of microbial contamination may include health consequences such as sickness, agony and suffering, starvation, and, in the worst-case scenario, death. Economic consequences for the affected person include pay loss, time away from work owing to illness, and healthcare-related costs. Time and money spent on consultations, treatment, investigations, and hospital stays, giving drugs and care have an impact on healthcare services. Figure 1.2 depicts a few instances of these contaminants.

1.4.1.2 Chemical Contaminant Candidates

Chemical contaminants are substances that are toxic to aquatic plants and animals. Chemicals are considered contaminants when they are present in places they should not be or in greater quantities than would naturally exist. Drinking water is also affected, which has serious health implications not just for humans but also for marine life and other organisms that consume contaminated water. These contaminants orig-

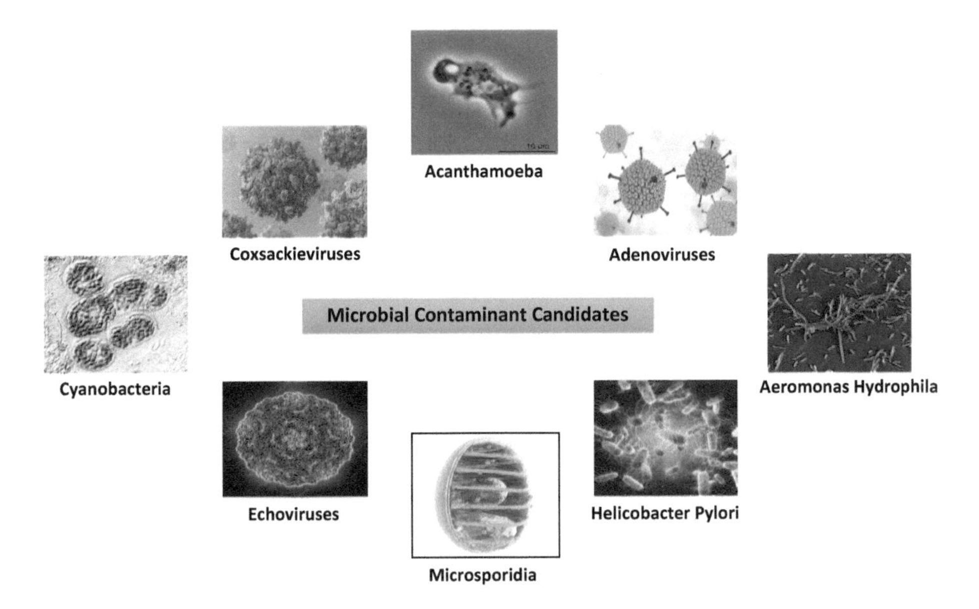

Figure 1.2: Selected examples of microbial contaminants.

inate in a variety of places, including industrial and municipal discharges, natural geological formations, urban and rural run-off, drinking water treatment processes, and water delivery materials.

Human activities such as hydraulic fracturing and horizontal drilling have also contaminated drinking water. Drinking water supplied from groundwater may also be contaminated with heavy metals such as nickel, mercury, copper, and chromium, potentially leading to an increase in cases of carcinogenic and noncarcinogenic health abnormalities. This type of source of drinking water contamination is very common in poor nations.

Chemical contaminants in drinking water include disinfection by-products, pharmaceutical by-products, pesticides, and other chemicals too. They have a wide range of health impacts, including cancer, cardiovascular disease, poor reproductive results, and neurological illnesses. Selected chemical contaminants are listed in Table 1.1, and selected chemical structures are shown in Figure 1.3.

1.4.1.3 Contaminants Levels

Long-term exposure to concentrations over the **m**aximum **c**ontaminant **l**evel **g**oal (**MCLG**, the highest contamination level in drinking water at which no known or predicted harmful effect on human health will occur, allowing for an adequate margin of safety) can result in liver damage, skin sores, appetite loss, blood and CNS issues, and skin sores. **M**aximum **c**ontaminant **l**evel (**MCL**, the highest level of a contaminant permitted in drinking water) are enforceable standards that must be set as closely as pos-

Table 1.1: Examples of chemical contaminants.

1,1,2,2-Tetrachloroethane	2,2-Dichloropropane	Bromobenzene
1,2,4-Trimethylbenzene	2,4-Dichlorophenol	DDE
1,1-Dichloroethane	2,4-Dinitrophenol	Diazinon
1,1-Dichloropropene	2,4-Dinitrotoluene	Dieldrin
1,2-Diphenylhydrazine	2,6-Dinitrotoluene	Disulfoton
1,3-Dichloropropane	*o*-Cresol	Diuron
1,3-Dichloropropene	Acetochlor	EPTC
2,4,6-Trichlorophenol	Boron	Fonofos
Linuron	Hexachlorobutadiene	Molinate
Manganese	*p*-Isopropyltoluene	Naphthalene
Methyl bromide	Metolachlor	Nitrobenzene
MTBE	Metribuzin	Organotins
Perchlorate	Vanadium	Terbacil
Prometon	Pesticides	Terbufos
RDX	Disinfection by-products	Triazines

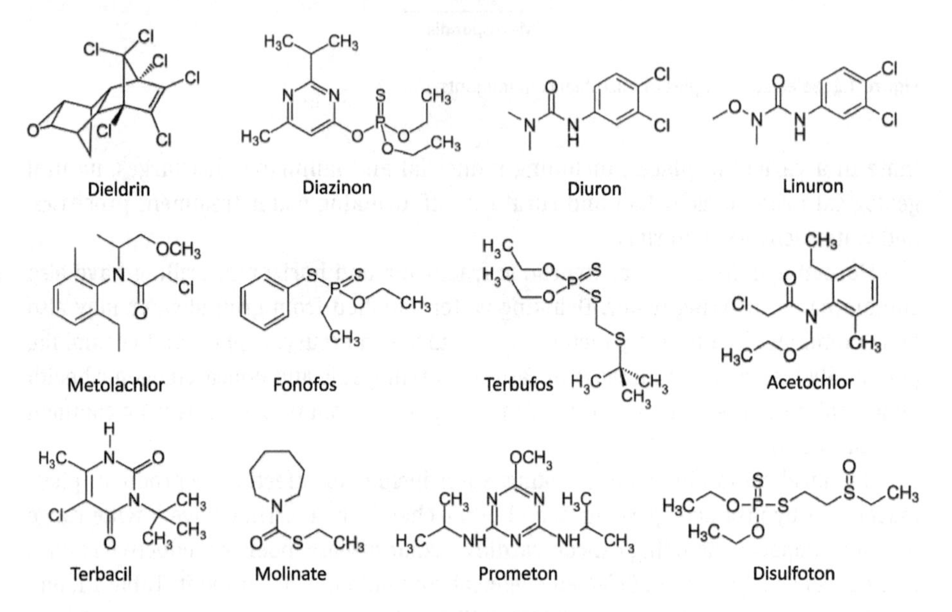

Figure 1.3: Chemical structures of selected chemical contaminants.

sible to the MCLG where the level of a contaminant in drinking water below which there is no known or anticipated risk to health. These standards are based on treatment technologies, affordability, and other factors that are feasible, such as the availability of analytical methods and treatment technology.

The examination of the negative effects brought on by the substance in question and the doses required to bring about such effects is the first step in establishing an

MCL. The end result of this procedure is a **ref**erence **d**ose (**RfD**), a safe dose based on research findings that have been extrapolated to humans from outcomes of animal tests.

MCLG is created for substances that don't cause cancer by first translating the safe dosage into a water concentration. Then, on the supposition that exposure to the chemical by drinking water represents just one-fifth of all possible exposure, this number is divided by five. The MCLG number is often the same as the MCL in most situations.

The MCLG is set at zero for substances that are thought to cause cancer, meaning that no level of the substance is thought to be acceptable. The MCL is based on the lowest concentration that can be consistently measured because 0 cannot be measured. The MCL is the lowest measured level for known or suspected carcinogens, not a safe level.

A number equivalent to the MCLG is determined for substances that are potentially carcinogenic, that is, there is some evidence that they might cause cancer, but this is not particularly compelling. The final MCLG is obtained by dividing this value by a factor of 10. This offers an additional level of safety in the event that the substance is ultimately shown to be carcinogenic.

1.5 Questions

1.5.1 What are the percentages of saline and fresh water in the world?
1.5.2 List three health benefits of drinking water.
1.5.3 Define chemical contaminants.
1.5.4 Give four examples of chemical contaminants.
1.5.5 Draw the chemical structures of 1,1,2,2-tetrachloroethane, 1,2,4-trimethylbenzene, and terbufos.
1.5.6 Give examples of disinfection by-products.
1.5.7 What does MCLG stand for?
1.5.8 Give three examples of microbial contaminants.
1.5.9 Define MCL.
1.5.10 What is the reference dose (RfD)?

Chapter 2
Water Chemistry

2.1 Properties of Water (Physical and Chemical Properties)

The earth is mostly made up of water (H_2O). Almost everything, including drinking, bathing, and cooking, requires water. Water makes up 60–70% of the human body. The survival of life on Earth depends on water. The earth's surface is unevenly covered in water. It dissolves practically all polar solutes and creates a significant solvent. At room temperature, water is a polar inorganic substance that is a tasteless, odorless liquid and is almost colorless. It is referred to as the "universal solvent" and the "solvent of life" and is by far the chemical compound that has been investigated the most. It is the most prevalent substance on Earth's surface and the only one that can be found as solid, liquid, and gas at the same time. Aside from carbon monoxide and molecular hydrogen, it is the third most prevalent molecule in the cosmos.

Water molecules are highly polar and form hydrogen bonds with one another. Because of its polarity, it can dissolve other polar substances such as alcohols and acids by interacting with them and dissociating the ions in salts. Its hydrogen bonding is responsible for several of its distinguishing properties, including a less dense solid form than its liquid form, a fairly high boiling point of 100 °C for its molar mass, and a large heat capacity. Water is amphoteric, which implies that it can exhibit acidic or basic properties depending on the pH of the solution. Because of its amphoteric nature, it undergoes a process of self-ionization. Figure 2.1 summarizes the unique features of water. Water molecules have multiple hydrogen bonds, which results in distinct properties when condensed. As a result, the melting and boiling points are both very high. Other liquids have lower specific heat, thermal conductivity, surface tension, and dipole moment than water. These qualities support its significance in the biosphere. Water assists in the transfer of ions and molecules required for metabolism. It has a high latent heat of vaporization, which helps regulate body temperature.

2.1.1 Density

The weight of water per unit volume, which fluctuates with temperature, is the definition of water's density. Grams per cubic centimeter, or 1 g/cm^3, is roughly how dense the water is. It responds to changes in temperature. Although liquid water, like other liquids, generally thickens as it is chilled at room temperature, pure water is thought to reach its maximum density at roughly 4 °C. It expands and loses density as the cooling process continues. This exceptional negative thermal expansion is the result of strong intermolecular interactions. Water lacks an absolute density due to its temperature-dependent density. It is denser in the liquid state than in the solid state. The

https://doi.org/10.1515/9783111332468-002

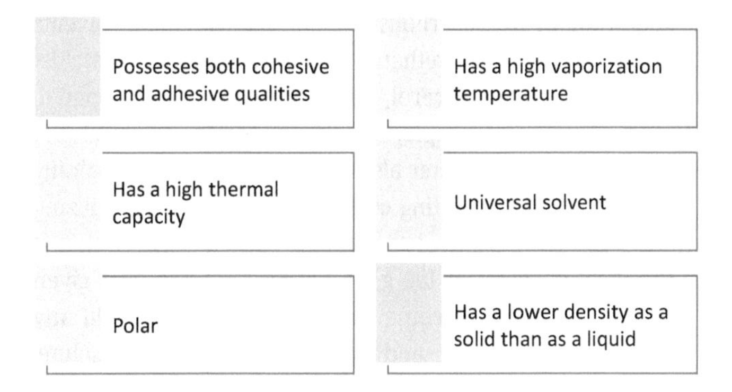

Figure 2.1: Unique features of water.

Table 2.1: The density of water for different temperature scale.

Temperature	Density (g/cm^3)
100 °C	0.9584
80 °C	0.9718
60 °C	0.9832
40 °C	0.9922
30 °C	0.9956
25 °C	0.9970
22 °C	0.9977
20 °C	0.9982
15 °C	0.9991
10 °C	0.9997
4 °C	0.9989
0 °C	0.9998
−10 °C	0.9981
−20 °C	0.9935
−30 °C	0.9838

density of water varies with temperature. The density (in g/cm^3) of water for various temperature ranges (from 100 to −30 °C) is given in Table 2.1.

2.1.2 Solubility

Water solubility is a measurement of how much a chemical compound may dissolve in water at a given temperature. Solubility is usually measured in mg/L (milligrams per liter) or ppm (parts per million). Intermolecular interactions determine a substance's solubility in a liquid, which also determines whether two liquids are miscible. Solutes are classed as hydrophilic (water-loving) or hydrophobic (water-fearing). For example, water is poorly soluble in aliphatic and aromatic hydrocarbons but miscible with polar solvents such as acetonitrile, dimethyl sulfoxide, dimethoxyethane, dimethylformamide, acetaldehyde, sulfolane, tetrahydrofuran, 1,4-dioxane, glycerol, acetone, isopropanol, propanol, ethanol, and methanol. Salt and sugar, for example, dissolve in water. Warm or hot water usually dissolves them faster and better. Water also provides a vital life-sustaining force that functions at the biological level by assisting cells in transporting and utilizing substances like oxygen or nutrients. Figure 2.2 depicts a simple solubility curve that is commonly used to calculate the mass of solute in 100 g (or 100 mL) of water at a given temperature. When it reaches the line, it has become saturated and cannot hold any more solute. Below the line indicates unsaturation and the ability to store more solute. Above the line, the solute concentration is higher than it should be.

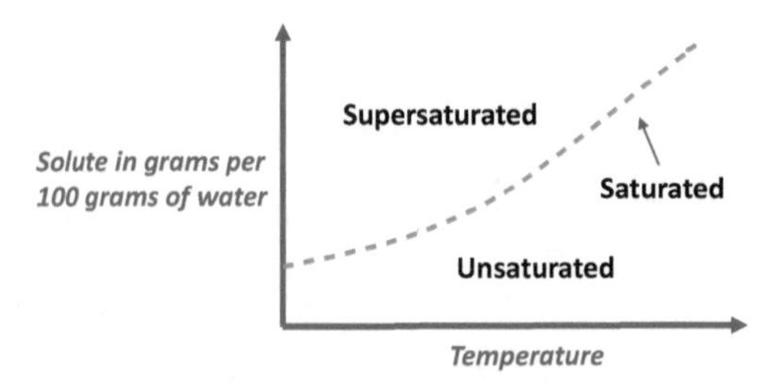

Figure 2.2: Simple water solubility curve.

2.1.3 Polarity

One of the most important characteristics of water is that it is a polar molecule, with the hydrogen and oxygen in water molecule (H_2O) forming polar covalent bond. The polarity of a water molecule results in a slightly positive charge on hydrogen and a slightly negative charge on oxygen despite the fact that a water molecule has no net charge. Because oxygen is more electronegative than hydrogen, a shared electron is more likely to be located there than close to the hydrogen nucleus, which results in a partial negative charge near the oxygen.

Due to the polarity of water, nearby water molecules are pulled to one another by their opposing charges and form hydrogen bonds. Water also attracts or repels other polar molecules and ions as shown in Figure 2.3. A polar chemical that easily interacts with or dissolves in water is said to be hydrophilic. Hydro- is the prefix for "water," and -philic is the suffix for "loving." Oils and fats, which are non-polar substances, do not mix well with water and separate from it rather than dissolve in it. These nonpolar substances are referred to as hydrophobic (hydro- = "water" and -phobic = "fearing") substances.

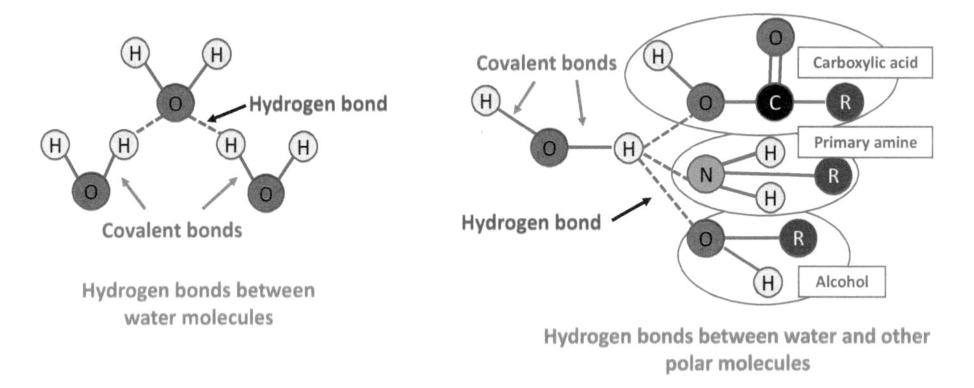

Figure 2.3: Hydrogen bond formation due to water polarity.

2.1.4 Acid–Base Behavior

When water can operate as both an acid and a base in solution, it can become both an acid and a base to itself during the autoionization process as shown in Figure 2.4. Water absorbs hydrogen ions and behaves as a base when combined with an acid. When water is combined with a base, it behaves like an acid because hydrogen ions are produced.

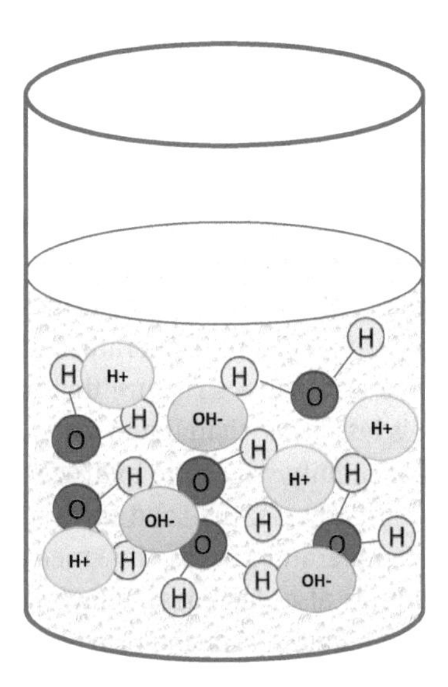

Figure 2.4: Water autoionization process.

The hydrogen (H^+) ions in acids are what cause their acidic nature. In the presence of water, they release hydrogen ions. The fact that water is a polar solvent aids in breaking down the link between the ions and making them soluble. On the other hand, when placed in water, bases draw hydrogen atoms. Sodium hydroxide, or NaOH, is an illustration of a base. Hydrogen ions are drawn to it when it is submerged in water, and as hydroxyl (OH^-) ions build up, a basic or alkaline solution is created. Figure 2.5 illustrates the amphoteric characteristic of water, which can function as both an acid and a base.

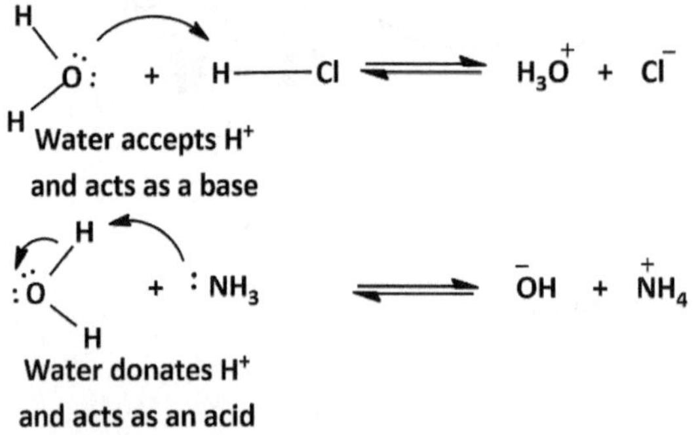

$$H_2O + H{-}Cl \rightleftharpoons H_3O^+ + Cl^-$$

Water accepts H^+ and acts as a base

$$H_2O + NH_3 \rightleftharpoons OH^- + NH_4^+$$

Water donates H^+ and acts as an acid

Figure 2.5: An illustration of the amphoteric characteristic of water.

2.1.5 Water States

The ability of liquid water to generate hydrogen bonds is a significant property that is essential to life. Since living organisms have a high-water content, understanding these chemical traits is essential to understanding life. As water molecules form hydrogen bonds with one another, the water acquires several distinct chemical properties compared to other liquids. As the water molecules flow by one another in liquid water, hydrogen bonds are continuously made and broken. These bonds are broken as a result of the water molecules moving because of the heat in the system, which gives them kinetic energy. The hydrogen bonds between water molecules entirely dissolve when the heat is increased during the boiling process, allowing water molecules to escape into the air as gas (steam or water vapor). However, as the temperature of water drops and it freezes, the water molecules create a crystalline structure that is sustained by hydrogen bonds, which results in ice that is less dense than liquid water, a phenomenon that is not observed when other liquids solidify. Since the hydrogen bonds between water molecules are forced away when it freezes, the solid form of water has a lower density than liquid water. When the temperature drops, the kinetic energy between molecules in the majority of other liquids also decreases, allowing the molecules to pack even more closely than they would in liquid form and providing the solid a higher density than the liquid. Because of its unusually low density, ice floats at the surface of liquid water, as seen in icebergs and ice cubes in ice water. Ice that accumulates on the surface of lakes and ponds functions as an insulator, keeping animals and plant life from freezing. Plants and animals in the pond would freeze in the solid block of ice and die if this layer of insulating ice was not present.

2.1.6 Water Redox Reaction (Water Redox Process)

When molecular hydrogen (H_2) is oxidized by molecular oxygen (O_2) to form water (H_2O), the reaction can be thought of as two coupled processes, as shown in Figure 2.6: electron transfer from hydrogen to oxygen (reduction of oxygen) and electron acceptance from hydrogen by oxygen (oxidation of hydrogen). The oxidizing agent is oxygen, and the reducing agent is hydrogen.

Water splitting is the process by which two molecules of water are broken down into their basic elements: two molecules of hydrogen gas (H_2) and one molecule of oxygen gas (O_2). The overall breakdown of water into oxygen and hydrogen is the same whether either half reaction pair is combined. The amount of hydrogen produced is double that of oxygen.

The importance of redox reactions in aqueous solutions in biological and environmental systems cannot be overstated. They support and sustain life by collecting and dispersing energy in order to develop and spread low-entropy living systems.

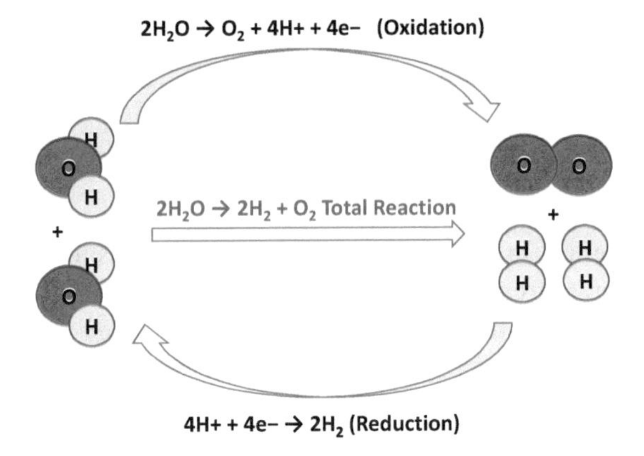

Figure 2.6: Water redox process.

2.1.7 Water's Cohesive and Adhesive Properties

When a glass of water is filled to the brim and then a few drips are carefully added till the glass overflows, the water forms a dome-like shape above the rim. This water can float over the glass due to its cohesion. In this case, water molecules are attracted to each other due to hydrogen bonding, which keeps the molecules together at the water-air interface.

Surface tension is usually promoted by cohesion. When water is dropped on a dry surface, it forms droplets rather than being flattened out by gravity. The paper floats despite the fact that paper is denser (heavier) than water when a small scrap of paper is placed on top of a droplet of water. Cohesion and surface tension keep water molecules' hydrogen bonds intact and the item afloat. If placed correctly and without breaking the surface tension, a needle can also "float" on top of a glass of water.

The adhesion property of water, or the attraction of water molecules to other molecules, is related to these cohesive forces. When water comes into touch with charged surfaces, such as those found on the interior of tiny glass tubes known as capillary tubes, this attraction can be stronger than the cohesive forces in the water. Figure 2.7 shows how water "climbs" up the tube submerged in a glass of water, giving the impression that the water is higher on the sides of the tube than in the center. This is due to the fact that water molecules stick to the charged glass walls of the capillary more so than they do to other molecules. This kind of adherence is referred to as capillary action.

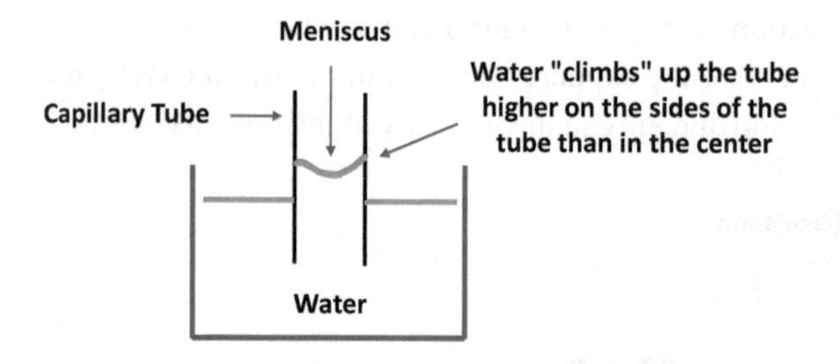

Figure 2.7: Water adhesion by capillary action.

Cohesive and adhesive forces also influence water transfer from a plant's roots to its leaves. Water molecules that evaporate from the plant's surface remain attached to water molecules below them, causing them to be dragged along. Plants employ this natural process to help move water from the roots to the leaves. Plants would be unable to receive the water and dissolved minerals they require if these water qualities were not present. Due to surface tension, aquatic insects such as water striders can also float on the water's surface.

2.2 Hard Water

Hard water contains a high concentration of mineral ions. The metal cations, calcium (Ca^{2+}) and magnesium (Mg^{2+}), are the most prevalent ions found in hard water, while iron (Fe^{2+}), aluminum (Al^{3+}), and manganese (Mn^{2+}) may also be present in some regions (Figure 2.8). These metals are water soluble, which means they dissolve in water. These ions' relatively large concentrations can saturate the solution, causing

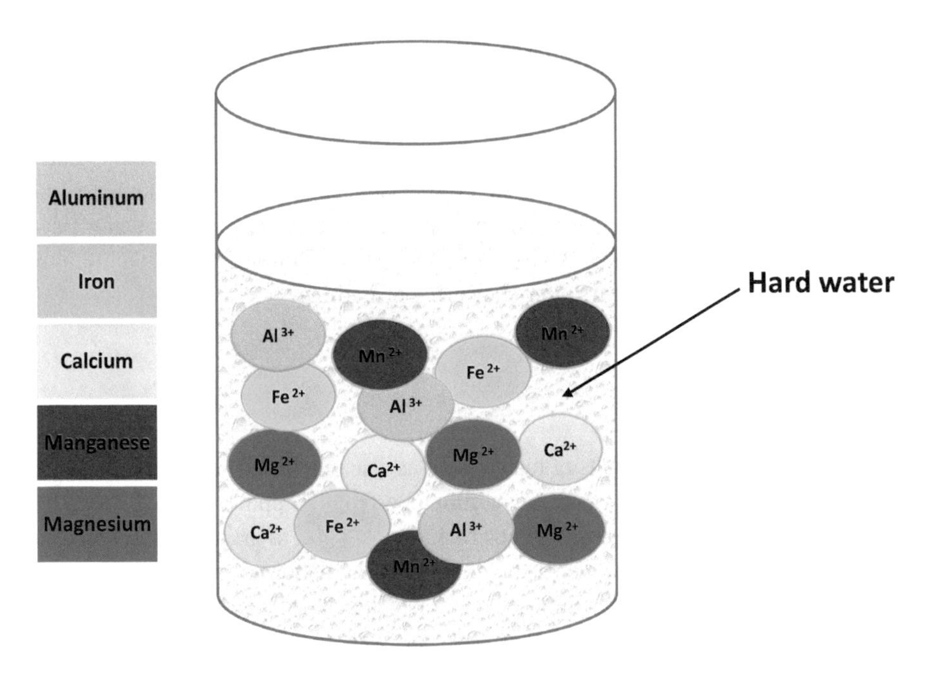

Figure 2.8: Hard water.

the equilibrium of these solutes to move to the left, toward reactants. In other words, the ions have the ability to precipitate out of the solution. The calcination, or precipitation of calcium or magnesium carbonate, visible on water faucets is caused by the displacement of minerals from the solution. Furthermore, hard water can react with other substances in a solution, such as soap, to form a precipitate known as "scum."

There are two forms of hard water: temporary and permanent. Temporary hard water is made up mostly of calcium (Ca^{2+}) and bicarbonate (HCO_3^-) ions. The bicarbonate ion in temporary hard water decomposes upon heating into carbonate ions (CO_3^{2-}), carbon dioxide (CO_2), and water (H_2O). The carbonate ion (CO_3^{2-}) formed can then combine with other ions in the solution to generate insoluble compounds such as $CaCO_3$ and $MgCO_3$. The interactions of carbonate ions in solution also generate the well-known mineral build-up visible on the sides of boiling pots, rust known as "boiler scale." The breakdown of the bicarbonate ion caused by increasing the temperature of transient hard water represents a shift in the equilibrium equation. This shift is responsible for the white scale seen in the boiling containers as well as the mineral deposits that form inside water pipes, resulting in inefficiency and, in extreme cases, explosion due to overheating. Because $CaCO_3$ or other scale is relatively insoluble, it does not entirely dissolve back into the water when cooled. As a result, this form of hard water is considered "temporary" since boiling can alleviate the hardness by removing the problematic ions from the solution.

Permanent hard water contains large levels of anions such as sulfate anion (SO_4^{2-}). This is referred to as "permanent" hard water because, unlike temporary hard water, the hardness cannot be erased merely by boiling the water and allowing the mineral ions to precipitate out. However, the name is misleading because "permanent" hard water can be softened by different methods. The permanent hard water scale has similar negative impacts as the transitory hard water scale, such as obstructing water flow in pipes. Permanent hard water is also to blame for the bathtub "ring," commonly known as soap scum, that appears after showering or bathing.

2.3 Soft Water

Soft water is free of dissolved salts of metals such as calcium, iron, and magnesium or it has a mineral content of calcium and magnesium of less than 17 ppm (17 ppm; Figure 2.9). It is primarily derived from peat or igneous rock sources such as granite, but it can also be derived from sandstone. Soft water contains very little dissolved salt. It is defined as water with less than 50 mg of calcium carbonate per liter when expressed in terms of a comparable amount of calcium carbonate. Most soft water has a pH of 6–7, while most hard water has a pH of 7–8. This is because hard water minerals lower the quantity of acid in the water; therefore when they are removed, the outcome is a lower pH level.

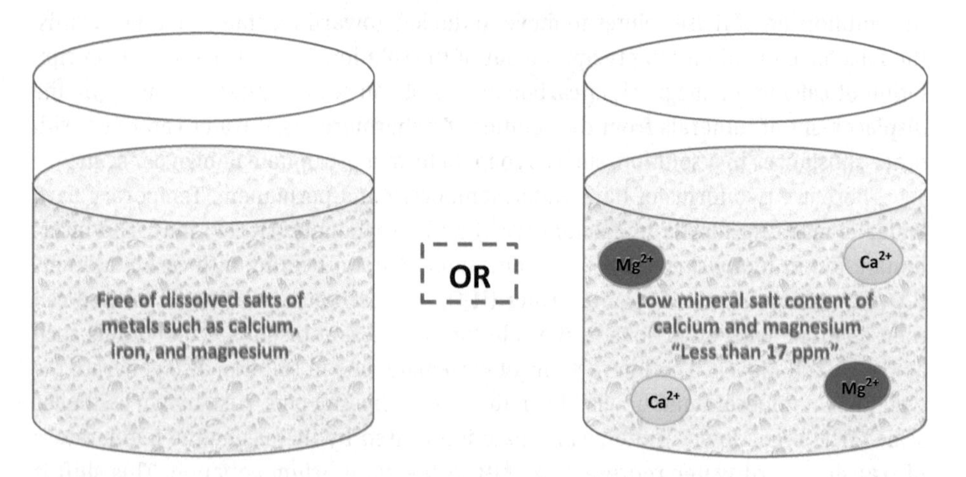

Figure 2.9: Soft water.

2.4 Heavy Water

Deuterium, an isotope of hydrogen, and oxygen are combined to form heavy water (also known as deuterium oxide or D_2O; Figure 2.10). The atomic weight of deuterium, which is 2, and the atomic weight of oxygen, which is 16, together make up the molecular weight of heavy water, which is approximately 20; contrast this with regular water, which has a molecular weight of about 18 (two times that of regular hydrogen, which is 1, plus oxygen, which is 16).

Heavy water differs from ordinary water in terms of its physical characteristics, such as being 10.6% denser and having a higher boiling point. At a given temperature, heavy water is less dissociated and doesn't have the same color as ordinary water. It doesn't taste much different, but it can taste a little sugary. When used as a coolant in nuclear reactors, it can become slightly radioactive even though it is not radioactive in its pure form. Deuterated water is naturally present in normal water, but it can be isolated via chemical exchange, electrolysis, or distillation techniques. The most practical way for producing heavy water from an economic standpoint is the Girdler sulfide process. Heavy water is sold and used in a range of industries and comes in various purity degrees. Its applications include neutron capture therapy, nuclear magnetic resonance, infrared spectroscopy, neutron moderation, neutrino detection, metabolic rate monitoring, and the production of radioactive elements like plutonium and tritium.

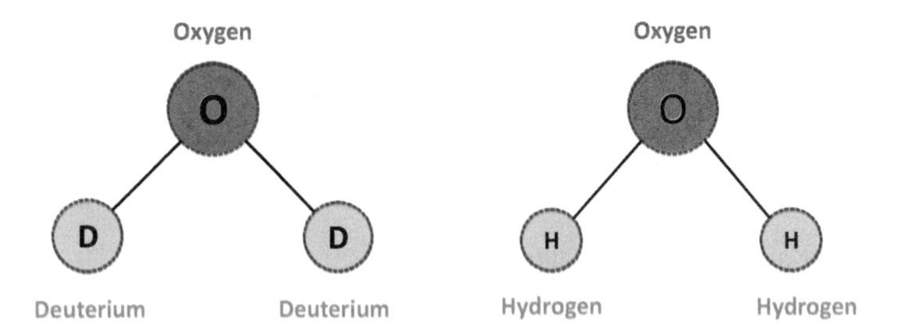

Figure 2.10: Heavy water (D_2O) versus water (H_2O).

2.5 Water Hydrates

Any chemical that contains water in the form of H_2O molecules is referred to as a hydrate. The most well-known hydrates are crystalline solids that decompose once the attached water is removed. Hygroscopic or deliquescent substances are those that spontaneously take up water from the air to create hydrates, whereas efflorescent hydrates are those that lose the so-called water of hydration or water of crystallization to become unhydrated (anhydrous) substances. When water is taken in and lost through heating, the color changes. As an illustration, the colors of blue vitriol (copper sulfate, $CuSO_4$), copper sulfate pentahydrate ($CuSO_4 \cdot 5H_2O$), and copper sulfate trihydrate ($CuSO_4 \cdot 3H_2O$) are all blue. Cobalt (II) chloride ($CoCl_2$) is another vivid illustration, and it can be used as a water indicator because it changes color from blue to red ($CoCl_2 \cdot 6H_2O$) when hydrated (Figure 2.11).

Blue anhydrous cobalt (II) chloride ($CoCl_2$) **Pink cobalt (II) chloride hexahydrate ($CoCl_2 \cdot 6H_2O$)**

Blue vitriol Copper sulfate ($CuSO_4$) **Blue copper sulfate pentahydrate ($CuSO_4 \cdot 5H_2O$)**

Figure 2.11: Examples of hydrated and dehydrated chemicals.

2.6 Questions

2.6.1 List three unique features of water.

2.6.2 What is the density of water at 0 °C and 100 °C, and is there a big difference in density?

2.6.3 Show how water forms hydrogen bonds with both amines and alcohols and why?

2.6.4 What does water autoionization mean?

2.6.5 Explain the amphoteric characteristic of water.

2.6.6 Give an example of a water redox reaction.

2.6.7 Why does water climb up the tube submerged in a glass of water higher on the sides of the tube than in the center?

2.6.8 List the most prevalent ions found in hard water.

2.6.9 What is the difference between temporary and permanent hard water?

2.6.10 Give three applications for heavy water.

Chapter 3
Water Sources

3.1 Natural Resources

Water resources are any of the many naturally occurring waters on the Earth, regardless of their state or ability to benefit humans. The most helpful water resources are those found in rivers, waterfalls, ponds, and lakes; other sources include deep underground water, glaciers, and permanent snowfields. The Earth's surface is covered by water over 71% of its area. Fresh water makes up just 3% of the world's total water supply, which makes up 97% of the seas. Glaciers and ice caps contain 2% of the world's fresh water. Selected examples of water resources are shown in Figures 3.1–3.4.

Figure 3.1: Almese Waterfall, Italy.

https://doi.org/10.1515/9783111332468-003

Figure 3.2: Como Lake, Italy.

Figure 3.3: Nova Ponente Pond, Italy.

Figure 3.4: Po River, Italy.

3.2 Water Cycle

The water cycle demonstrates how water is constantly moving both inside the Earth and in the atmosphere. It is a complicated system with a wide range of processes. Water vapor is created when liquid water evaporates, and this water vapor then condenses to form clouds and falls as rain and snow back to the Earth. The three phases of water – atmospheric water, surface water, and groundwater – are in constant circulation. The moisture that clouds are made of is known as atmospheric water. Warm air pushes the vapor upward into colder levels, where it condenses to form clouds and returns to the Earth as rain or snow. The water found in rivers, seas, and oceans is considered surface water. Deep underground water is known as groundwater. The water goes in three different directions: some move to surface bodies, some move deeper through interstices (gaps or holes), and the remainder moves deeper through impervious strata before becoming groundwater.

The water cycle has four primary stages: Figure 3.5 illustrates them as evaporation, condensation, precipitation, and collection.

- Evaporation is the process through which water from lakes, oceans, streams, ice, and soil rises into the air and evaporates to form water vapor and then water vapor droplets combine together to form clouds.
- Condensation is the process through which airborne water vapor cools and transforms back into liquid water.
- Precipitation is the process by which water from clouds in the sky descends to the ground as rain, snow, hail, or sleet.
- Collection is the process that takes place when water from the clouds gathers in the oceans, rivers, lakes, and streams as rain, snow, hail, or sleet. The majority will permeate (soak into) the earth and gather as underground water.

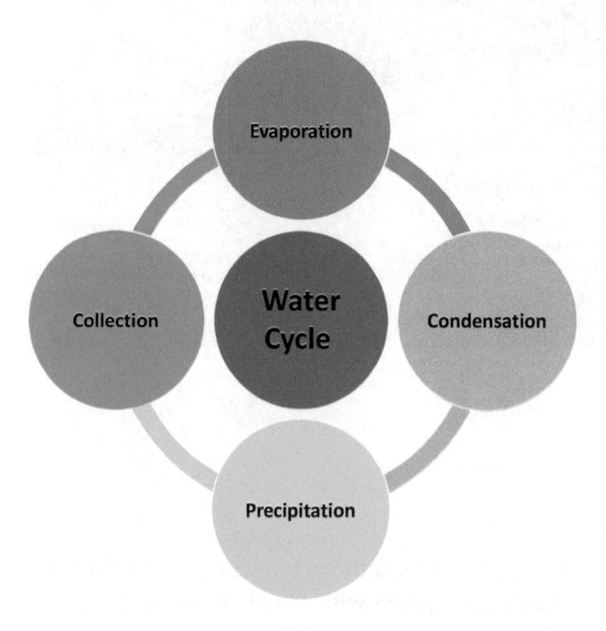

Figure 3.5: Water cycle.

3.3 Water Supply

There are three separate natural water sources in total. They are classified into three types: surface water, groundwater, and rainwater. Snow and other types of precipitation are also sources of rainwater. Precipitation that falls on the Earth is important in the water cycle because it replaces both surface and underground water supplies. These are the water sources that you cannot see or easily access. Underground water sources are critical to the water cycle since they are the primary sources for human use. Wells are typically utilized to access these underground water sources, and springs are employed to force the water to the surface when the pressure is too high. Surface water is the easiest to see and use. The negative is that most of the world's surface water is salt water, which is unsuitable for drinking by most living things. In addition to serving as a source of drinking water, surface water plays a vital role in our everyday life. Surface water is utilized to generate hydroelectric power, which is a clean and renewable energy source. Precipitation, springs, and ice melting from higher elevations and glaciers supply surface water. Probably the world's largest source of water is "hidden" or trapped in a different type of water. Hidden sources of water can be found in glaciers, polar ice caps, and even the atmosphere. Although these sources are not immediately usable, they all eventually become fresh water due to melting and precipitation.

3.3.1 Surface Water

Surface water is defined as any body of water found on the Earth's surface, including saltwater in the oceans and fresh water in waterfalls, rivers, and lakes. Precipitation that does not soak into the ground or evaporate is referred to as runoff. There are three forms of surface water: perennial, ephemeral, and man-made. When there is little or no precipitation, perennial or permanent surface water remains throughout the year and is recharged by groundwater. Ephemeral or semipermanent surface water exists only for a portion of the year. Ephemeral surface water includes small creeks, lagoons, and water holes. The third form is the man-made canals and lakes. Selected examples of the three forms of surface water are depicted in Figures 3.6–3.13.

Figure 3.6: Indian Ocean, Mauritius.

Figure 3.7: Niagara Falls, Canada.

Surface waters are divided into two types: lentic and lotic. The lentic zone is described as the area of a water body where there is no flow and the water is stagnant, whereas the lotic region is defined as the area of a water body where the water is in a continuous state of motion, that is, in a dynamic state.

3.3.1.1 Lentic Water

Lentic water or lentic ecosystems refer to aquatic systems that move so slowly that the water seems static or nearly so. Ponds and lakes are some examples. Algae-rooted and floating-leaved plants and crustaceans like crabs and prawns live in these environments. Amphibians such as frogs and salamanders, as well as reptiles such as alligators and water snakes, can be found here. During droughts, complex systems frequently outlast their smaller counterparts, allowing species to survive despite limited resources. Many things happen in these bodies, which are often deep and have multiple zones, that lotic water systems do not. Figure 3.14 depicts the littoral, limnetic, profundal, and benthic zones of the lentic water.

The littoral zone refers to the shoreline of a lake or pond. It includes everything from the dry land to the water's edge along the coast or bank. Depending on the location and age of the lake, the extent and depth of this zone might vary substantially.

Figure 3.8: Geneva Lake, Switzerland.

However, this zone is often shallow and rich in nutrients from runoff and inbound water sources. As a result, it is the zone with the most aquatic or semi-aquatic flora, such as reeds, grasses, and algae.

The next layer is the limnetic zone, which is the lake's surface or open water area. The amount of light that enters the body of water defines the limnetic zone. This upper water layer, also known as the euphotic zone, is the warmest and receives the most sunlight in the lake. The zone terminates when sunlight can no longer penetrate the lake. Because of the presence of sunlight, aquatic plants grow in this region, as they do in the littoral zone. Because oxygen levels are greater in this region of the lake, the majority of fish also reside there.

The profundal zone starts when sunlight can no longer reach the lake's surface. The temperature in these oceans is considerably lower as a result of the Sun's heat not being able to penetrate these depths. The profundal layer's volume and depth will differ from lake to lake since the purity and make-up of the water greatly affect how far sunlight may travel. There are much fewer fish in this area of the lake since there is less oxygen there.

Figure 3.9: Columbia Lake Waterloo, Canada.

The region along the lake's bottom is known as the benthic zone. The sediment, silt, and dirt that accumulate at the lake's bottom are all included in what makes up the bottom of the body of water. Bacteria exist and work to break down any organic material that has fallen to the lake floor at this lowest position; everything, including dead vegetation, fish, animals, and animals' droppings. Benthic zones are larger or more prevalent in older lakes because there is more material to decay.

Thermal stratification is a phenomenon that causes some lakes and ponds in temperate regions to divide into three distinct thermal layers or zones: the epilimnion, metalimnion, and hypolimnion (Figure 3.15).

In the epilimnion or surface water, is where life is most abundant. The warmer region that receives the most sunshine will hold the most oxygen, allowing the most life to flourish here. This zone has a high oxygen concentration during the summer.

The temperature of the water starts to drop in the metalimnion or middle mass of the lake. Although there are not as many species here as in the surface waters, there are still many to be discovered. The thermocline, which is a region where the water temperature drops by around one degree Celsius each meter, frequently starts here.

The hypolimnion is the zone with the greatest temperature differences between summer and winter. It is the area of the water body closest to the bottom where light may not always penetrate. Most of the year, the least quantity of life can be found in

Figure 3.10: Sirio Lake, Italy.

this zone. When water is stratified in bodies, overturning happens when the water near the surface and the water at the bottom of the mass mix. This happens in the spring and autumn.

The critical topic is how to keep lentic ecosystems safe and safeguarded against threats that can harm the entire ecosystem. Many of these hazards are caused by human intervention, while others are caused by climate change. Figure 3.16 summarizes some of these threats.

3.3.1.2 Lotic Water

Lotic water, flowing waterways, include rivers, streams, and springs. Because of their steady movement their oxygen concentration is higher, and their water is clearer. Lotic ecosystems are home to a variety of insect species, including beetles, stoneflies, and mayflies, all of which have acquired environmental adaptations. A variety of fish species, including eels, trout, and minnows, can be found here. A variety of species live in these environments, including beavers, otters, and river dolphins.

Lotic water systems are flowing continuously, ranging from torrential rapids to slow-moving backwaters and their water tend to be much shallower than their counterparts, causing the temperature to become a major abiotic factor for life. Water

Figure 3.11: Lachine Canal, Montreal, Canada.

found in these systems will freeze much quicker and thaw much faster than the deep waters of lentic systems. Lotic ecosystems depend on precipitation, snow melt, and springs to keep the water flowing. In times of drought these shallow systems will dry up and many organisms will die. Lotic environments have water flowing in only one direction, which is also known as unidirectional (one-way). As a result, there is a condition of constant physical change. Flowing rivers can sculpt the streambed through erosion and deposition, resulting in a variety of ecosystems such as glides and pools.

Light is required by biotic systems to provide the energy required to fuel the primary generation of new organic matter via photosynthesis. Prey animals can find refuge in the shadows cast by light.

All aerobic organisms require oxygen to survive in lotic ecosystems. The majority of oxygen enters the water via diffusion at the water-air interface. Its solubility in water diminishes as the temperature of the water rises.

The geologic material present in the catchment that is eroded, transported, sorted, and deposited by the current forms the underlying layer of inorganic substrate in lotic ecosystems. Boulders, pebbles, gravel, sand, and silt are examples of inorganic sub-

Figure 3.12: Venice Canal, Italy.

strates. The organic substrate can also comprise small particles, autumn fallen leaves, buried wood, moss, and more advanced plants.

Lotic water productivity and biodiversity are diminishing globally due to climate and anthropogenic-caused changes. Many current and developing challenges have an impact on their quality and availability. Table 3.1 offers a number of examples of the threats to lotic ecosystems.

3.3.2 Groundwater

After soaking into the soil as a result of rain or the melting of ice and snow groundwater is preserved in small holes between rocks and soil particles. Although it is chilly, transparent, and odorless and has the least turbidity of any natural water, it is quickly contaminated by limestone deposits or septic tank effluent.

The movement of groundwater is a component of the hydrologic cycle. As precipitation and other surface water sources recharge it, the groundwater gradually – and occasionally very gradually – drains toward its discharge point. Groundwater does not

Figure 3.13: Waterloo Park Silver Lake, Canada.

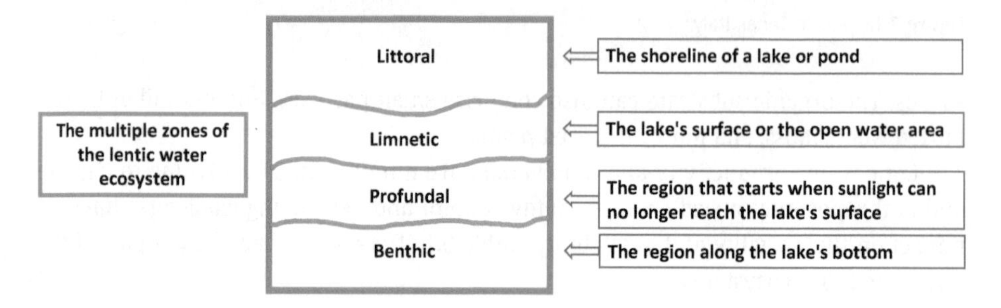

Figure 3.14: Lentic water ecosystem multiple zones.

wait to be drawn from a well or remain motionless indefinitely beneath the surface. Groundwater is present in two zones. The pores, or open spaces, that make up the unsaturated zone, which is located just below the Earth's surface, contain both water and air. The saturated zone, which is located underneath the unsaturated zone and is where all the pores and rock cracks are filled with water, is shown in Figure 3.17.

Numerous factors can lead to groundwater contamination. An aquifer's recharge surface water will contaminate the groundwater if it is tainted. The quality of surface water at discharge points may then be impacted by contaminated groundwater. Liquid hazard-

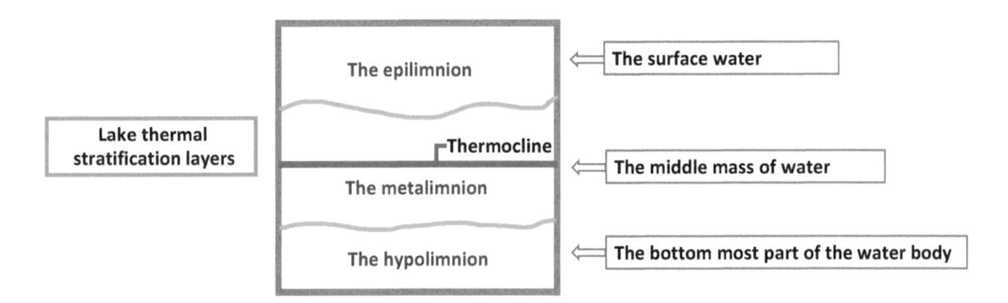

Figure 3.15: Lake thermal stratification layers or zones.

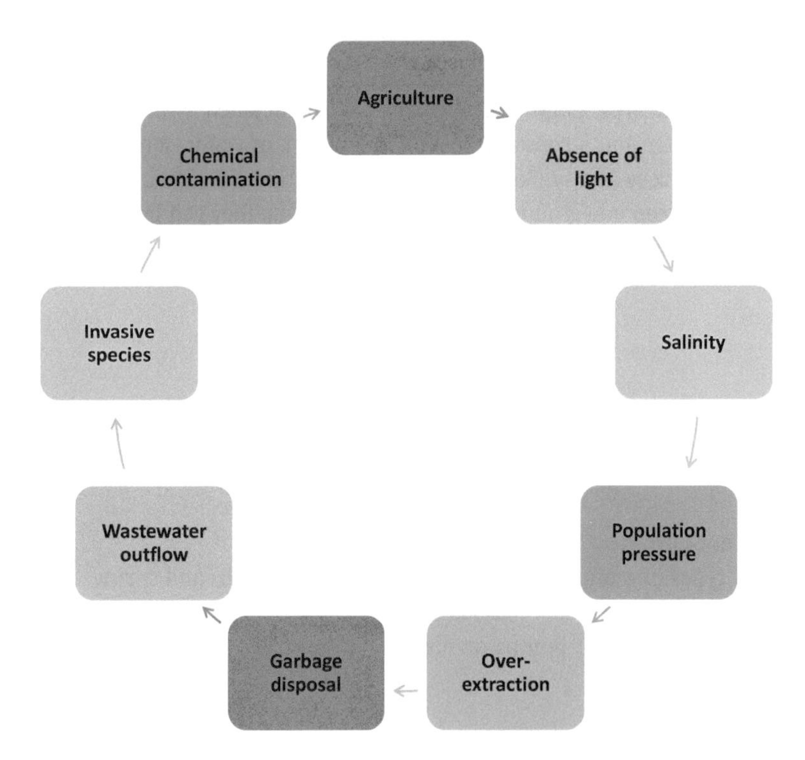

Figure 3.16: Threats to lentic ecosystems.

Table 3.1: Threats to lotic ecosystems with examples.

Industrial effluents	DDT, dyes, mercury, cadmium, lead. Solid wastes such as metals, plastics, and artificial fibers
Agrochemicals	Fertilizers, herbicides, and pesticides
Heavy metals	Mercury, lead, cadmium, copper, zinc, and nickel

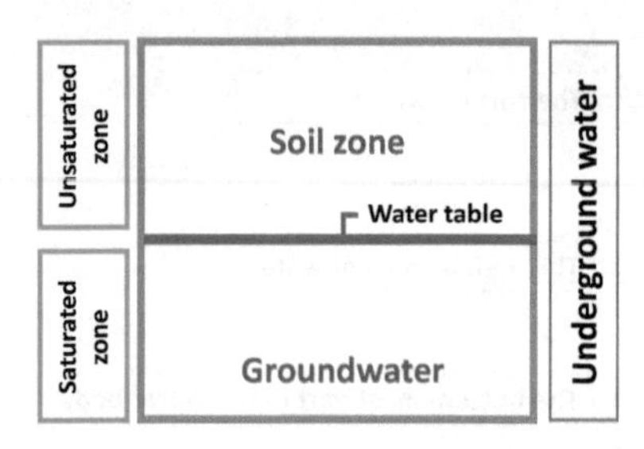

Figure 3.17: Groundwater zones.

ous compounds that seep into groundwater through the earth can also contaminate it. The water will carry contaminants that can dissolve in groundwater with it, possibly to wells that provide drinking water. An area of contaminated groundwater, known as a plume, can occur if there is a persistent source of pollution entering flowing groundwater.

Very large volumes and areas of groundwater can get contaminated by the combination of moving groundwater and a constant source of pollution. In water, some dangerous compounds dissolve very slowly. Some of the contaminants will remain in liquid form if these compounds seep into groundwater more quickly than they can break down. The liquid will float on top of the water table like oil on water if it is less dense than water. In this state, pollutants are referred to as **l**ight **n**on**a**queous **p**hase **l**iquids **(LNAPLs)**.

The pollutants are referred to as **d**ense **n**on**a**queous **p**hase **l**iquids **(DNAPLs)** if the liquid is denser than water. DNAPLs develop pools at the aquifer's bottom as they sink. As the pools slowly decompose and are removed by flowing groundwater, they continue to poison the aquifer. Tiny globules of liquid are caught in the gaps between soil particles as DNAPLs move downward through an aquifer.

3.3.2.1 Groundwater Sources

The most prevalent types of groundwater sources are hand-dug wells and springs. Depending on the source type, multiple methods are used to abstract groundwater. Traditional techniques, such as hand pumps or mechanical or electrical pumps, can be

Table 3.2: Groundwater sources.

Spring	A spring is a location where underground water finds its way to the land surface and emerges, sometimes as a trickle, sometimes immediately after a rain, and sometimes in a constant flow. Hot springs can also arise when spring water emerges from subsurface heated rock. The quantity and velocity of the spring are both affected by the size of the aquifer and the location of the spring. The flow rate increases with the size of the aquifer.
Well	A well is a hole bored into the ground to gain access to water in an aquifer. To extract water from the ground, a pipe and a pump are needed, while a screen screens out undesired particles that could clog the pipe. Deeper wells have lower turbidity, more dissolved minerals, and fewer microorganisms. Natural filtration is less effective in shallow wells. Wells can be safeguarded by keeping them away from human waste and from geological leaks.

used to extract groundwater from boreholes and hand-dug wells. The two sources of groundwater are listed and discussed briefly in Table 3.2.

Wells are classified into four types. These types are briefly presented in Table 3.3.

Table 3.3: Well types.

Drilled well	A drilled well consists of a hole that has been excavated (also called a borehole) and a casing that lines the upper part of the entire depth of the well. The borehole is created by using a portable drilling machine.	 https://wellowner.org/2021/03/some-common-mistakes-well-owners-make-regarding-their-wells-and-equipment/
Bored well	Is built with an earth auger machine and 30 m deep. The well, which is normally half a meter in diameter, is constructed from concrete. This shallow well is more susceptible to pollution since it is the first to dry up during droughts.	 https://water.ca.uky.edu/boredwells
Dug well	Is not the ideal choice for the purpose of providing drinking water because it is shallow, often less than 15 m deep and 1 m broad. Due to its inadequate protection from surface water, it poses the largest risk of contaminating the water supply. A dug well can be built by hand or with the aid of machinery for excavation.	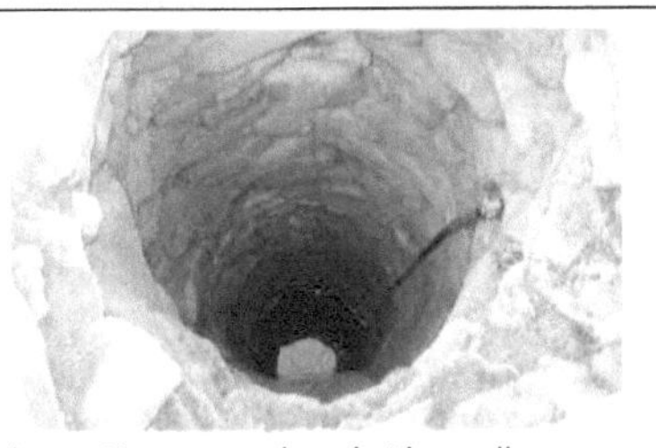 https://water.ca.uky.edu/dugwell
Driven or sand point well	The combined lengths of pipe are driven into the ground to create the sand-point or driven-point wells. This well typically has a diameter of no more than 5 cm and a depth of no more than 15 m. Only areas with relatively loose soils, such as sand and gravel aquifers, can accommodate this type of well. It is only put in locations with a shallow water table and few to no stones. It is driven into the ground or inserted using high water pressure.	 https://dengarden.com/home-improvement/How-To-Replace-Your-Old-Well-And-Save-Dollars

The wells mentioned above do not produce water in a sustainable way, and their water yield can decrease after being continuously drained for a long time. Figures 3.18 and 3.19 provide an overview of the causes of poor yield and the corrective actions that should be taken to fix them so that a reasonable yield is maintained.

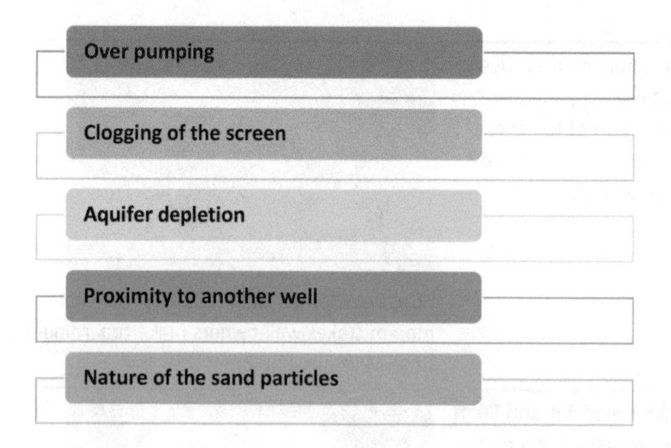

Figure 3.18: Causes of poor well yield.

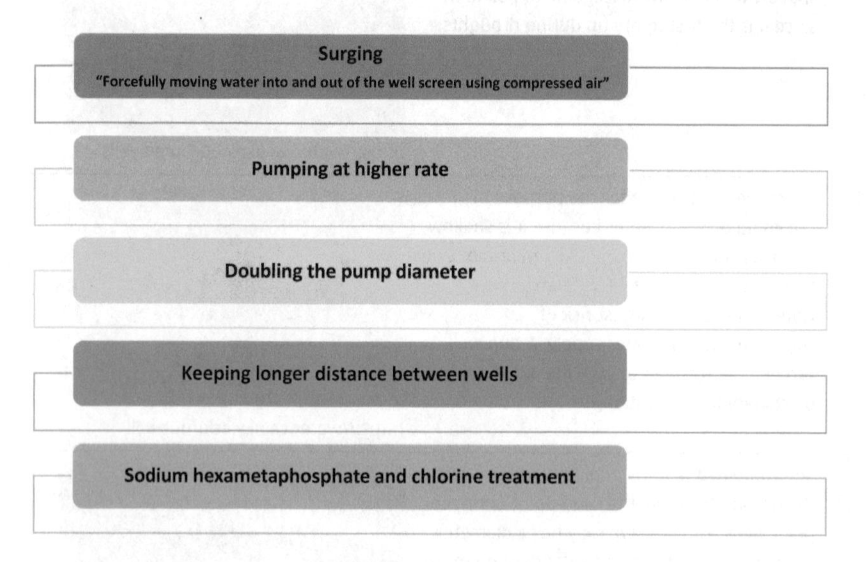

Figure 3.19: Well lower yields corrective actions.

3.3.3 Rainwater

Rain occurs when water vapor in the atmosphere condenses into droplets heavy enough to fall to the earth. Rain, which is an important aspect of the water cycle, deposits the bulk of the fresh water on the Earth. Given the right precautions, it is a typically safe source of water that can even be used for drinking. It is abundant in roof catchments and other paved places, and it can be used for a variety of purposes, including drinking, washing clothes, flushing toilets, watering gardens, and cooling and heating systems.

Rainwater often lacks any dissolved particles but contains dissolved gases such as carbon oxides, nitrogen, and sulfur, resulting in pH levels of 5.5 or below. Rainwater in coastal locations may include up to 15 mg/L of sodium chloride from sea spray. Rainwater also lacks alkalinity; it is acidic (low pH), low in mineral content, and aggressive to calcium-containing materials such as concrete and some metals commonly used in household plumbing. Rainwater is a crucial source of fresh water that is required for human, plant, and animal survival. Rainwater fills wells, lakes, and rivers, providing fresh water for home use.

Rainwater can be collected and saved for later use. Rainwater harvesting systems must include a collection area, a conveyance system, and a storage tank. A pump, a water treatment system, and water using fixtures can also be included in the systems. This type of water is often used as a non-potable supply for purposes such as toilet flushing and irrigation. It can also be used as potable water for drinking, dishwashing, and bathing if treated. Because treatment can be costly, most rainwater collection systems in metropolitan areas are not designed to provide drinking water.

Using rainwater to meet water supply needs helps to conserve groundwater in areas where water is supplied by groundwater sources, saves money and energy associated with the reduced use of treated municipal water, reduces the volume of stormwater runoff, which helps to prevent flooding and channel erosion, and saves money associated with drainage infrastructure and downstream stormwater management.

3.4 Questions

3.4.1 Give three examples of water resources.
3.4.2 What are the water cycle's four primary stages?
3.4.3 What are the three forms of surface water?
3.4.4 Define lentic water and give two examples.
3.4.5 Name the multiple zones of the lentic water.
3.4.6 Describe the thermal stratification phenomenon.
3.4.7 List five threats to lentic ecosystems.
3.4.8 Give examples of the threats to aquatic ecosystems.
3.4.9 What are the groundwater zones?
3.4.10 What are the causes of poor well yield?

Chapter 4
Water Pollution and Pollutants

4.1 Water Pollution

Water pollution occurs when dangerous substances, most typically chemicals or microbes, pollute a stream, river, lake, ocean, aquifer, or other water bodies, lowering water quality and making it poisonous to humans or the environment.

Water contamination is a major problem that endangers human health. Every year, unsafe water kills more people than war and all other types of violence combined. Meanwhile, our available drinking water is limited. We have access to less than one percent of the world's fresh water. Without action, the difficulties would only worsen when global demand for fresh water exceeds current levels. The five consequences of water pollution are shown in Figure 4.1.

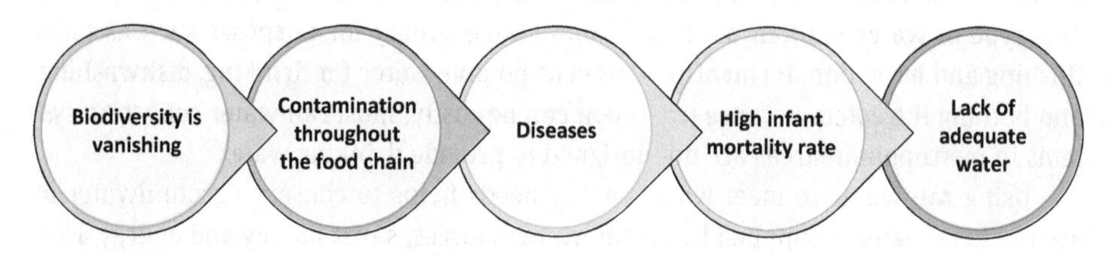

Figure 4.1: The five consequences of water pollution.

4.2 Water Pollution Causes

Water is very vulnerable to pollution. Water, widely described as a "universal solvent," has the ability to dissolve more compounds than any other liquid on the Earth. It is responsible for the vivid blue waterfalls. It also explains the ease with which water can be polluted. Toxic compounds released by farms, cities, and industry dissolve and combine with it, contaminating the water. Here are some of the most prominent global water pollution sources.

4.2.1 Agricultural Sector

Agriculture, the world's largest user of fresh water, contributes significantly to environmental pollution. It is considered the world's leading source of water pollution. Agricultural pollution is the leading cause of pollution in wetlands, lakes, rivers, and streams and also contributes to groundwater contamination.

https://doi.org/10.1515/9783111332468-004

Pesticides and fertilizers are two of the most common agricultural pollutants. The goal of these fertilizers and pesticides is to kill the local pests that cause crop harm. Some pesticides are absorbed by the soil, water, and plants when they are sprayed on crops. These chemicals are extremely dangerous since they affect the water supply and soil. Animals' health is harmed when they consume these leftover crops after harvesting. When we consume these crops, they may have an effect on our health.

Heavy metals like cadmium, lead, and arsenic can be found in insecticides and fertilizers. These heavy metal traces disintegrate in the soil, mix with water sources, are sucked up by groundwater, and are sometimes absorbed by crops.

Although agricultural manure and fertilizers may not be immediately harmful, they may include high levels of chemical nutrients such as nitrogen and phosphorus. These are the major sources of nutrient contamination that causes agricultural pollution, and their overabundance causes a threat to the quality of the world's water supply.

Non-native crop development and a lack of natural species are two other major contributors to agricultural pollution. This has an impact on human health and causes irreversible alterations in the natural environment. These plants may consume more soil nutrients and may emit toxins that harm the soil and water. As a result, crop output and quality may decline.

4.2.2 Sewage and Wastewater

While the term "sewage" is occasionally used to refer to all wastewater, sewage is the fraction of wastewater that contains feces or pee. It comes from the sinks, showers, and toilets in our homes. Wastewater is a catch-all term for the liquid waste that comes from industrial, commercial, and agricultural operations. Stormwater runoff, which occurs when rain causes impermeable surfaces to leak chemicals, oil, grease, and debris into our waterways, is another example of the phrase.

Wastewater is harmful to the environment because it carries germs, viruses, and disease-causing organisms that can taint shellfish populations, damage beaches, and prevent people from enjoying themselves or drinking water or eating shellfish.

In addition to polluting waterways with excess nutrients, sewage discharges also cause fish kills and coral reef die-offs by fostering hazardous algal blooms that endanger human health. Because sewage treatment is expensive, it is sometimes discarded without being cleaned. When sewage is dumped, it spreads bacteria, worms, or dangerous chemicals that can cause diseases including cholera, typhoid, and hepatitis. Sewage spills can often result in a variety of issues, some of which are shown in Figure 4.2.

Large amounts of wastewater from manufacturing facilities, oil refineries, wastewater treatment facilities, and septic systems flow back into the environment either legally or illegally without being treated or reused. If put in place and used correctly, wastewater treatment facilities can process a significant portion of wastewater. Prior to releasing the cleaned waters back into waterways, these facilities filter out pollu-

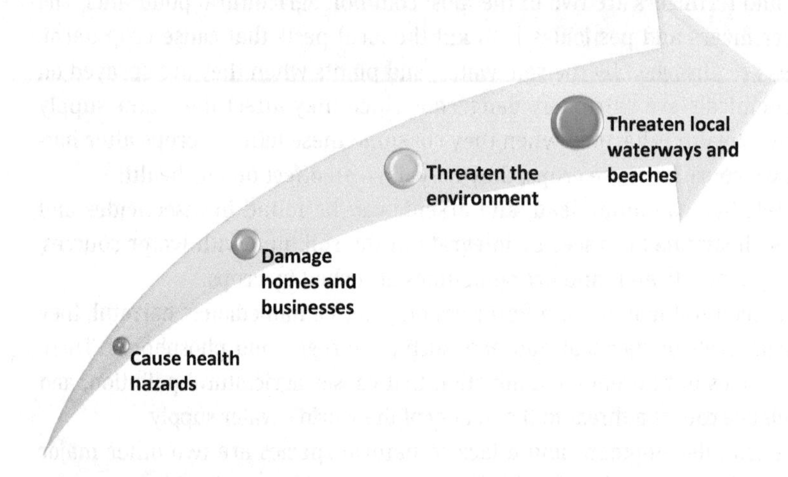

Figure 4.2: Sewage spill effects.

tants like bacteria, phosphorous, and nitrogen from sewage as well as heavy metals and hazardous chemicals from industrial waste.

Although wastewater treatment generates energy, fertilizers, lowers hazards to the public's health, and cleans up the water, it also emits a strong odor, hybrid microorganisms, needs a lot of money and space, and has an impact on the environment.

4.2.3 Oil Spills

Large portions of our economy are powered by oil, which is also used to heat our houses and provide energy. Accidental oil spills into the ocean or sea can injure marine life, sour a day at the beach, and render fish unfit for consumption. To remove the oil, assess the effects of pollution, and aid in the ocean's or sea's recovery, strong science is required.

Anywhere that oil is drilled, transported, or used is subject to oil leaks. The great majority of oil that enters marine habitats each year originates from sources on land as well, including factories, farms, and towns, in addition to tanker disasters. Different techniques are used to clean up oil spills. Figure 4.3 presents a summary of these techniques.

4.2.4 Radioactive Waste

Radioactive waste is defined as any pollutant that emits more radiation than the environment normally does. It is created by uranium mining, nuclear power plants, the de-

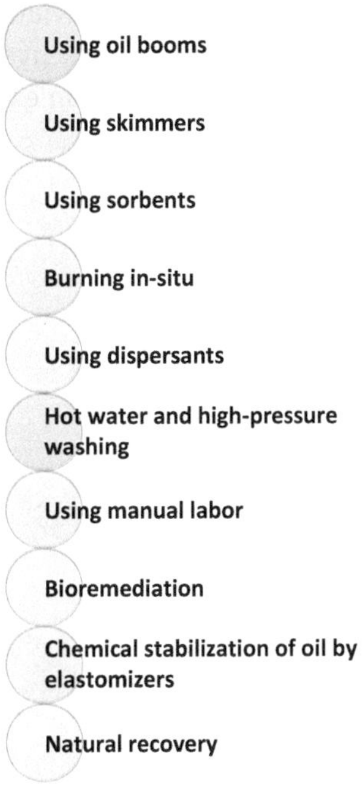

Using oil booms

Using skimmers

Using sorbents

Burning in-situ

Using dispersants

Hot water and high-pressure washing

Using manual labor

Bioremediation

Chemical stabilization of oil by elastomizers

Natural recovery

Figure 4.3: Oil spill cleaning methods.

velopment and testing of military weapons, and the use of radioactive materials in academic institutions and healthcare facilities' research and treatment plans. Because radioactive waste can last for tens of thousands of years in the environment, disposal is extremely challenging. Contaminants pose a risk to groundwater, surface water, and marine resources if they are unintentionally released or disposed of improperly.

Nuclear waste is classified into three types: high-level waste, transuranic waste, and low-level waste, and each must be disposed of according to the risk to human health and the environment. Some examples include radioactively contaminated protective shoe coverings and clothing, cleaning rags, mops, filters, reactor water treatment residues, equipment and instruments, medical tubes, swabs, hypodermic syringes, and laboratory animal corpses and tissues.

Radioactive waste can cause DNA damage in human cells. **A**cute **r**adiation **s**yndrome (**ARS**) or **C**utaneous **r**adiation **i**njuries (**CRI**) can be caused by high doses of radiation. High radiation exposure may potentially cause cancer later in life.

Low-level waste disposal is simple and may be done safely and practically in any place. Deep geological disposal is largely accepted as the best approach for the final disposal of the majority of radioactive waste produced. Radioactive materials are handled in special containers. Only non-life-threatening amounts of radioactive material are transported in type A packaging. Type B packaging is intended for transporting radioactive materials with the greatest levels of radioactivity.

Radioactive materials have the potential to infiltrate the human body if individuals breathe them in or consume contaminated food or water. Radiation or radioactive waste overexposure may have detrimental repercussions. Figure 4.4 illustrates an example of each.

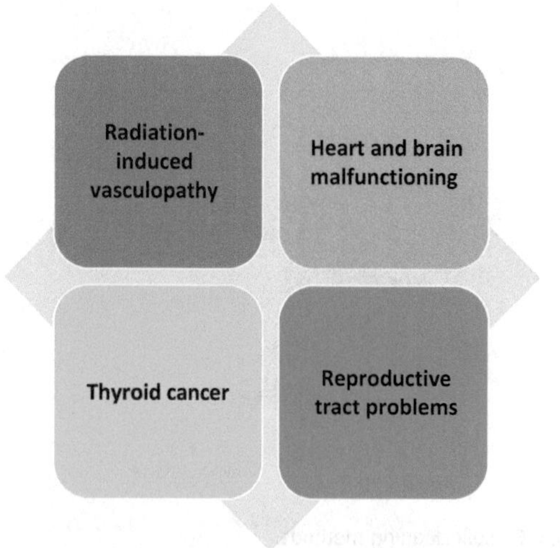

Figure 4.4: The effects of being exposed to radiation or radioactive waste.

4.3 Pollution Impacted Water Type

4.3.1 Groundwater Pollution

When rain falls, it seeps deep into the earth, filling fissures, crevices, and porous places in an aquifer to form groundwater. It is one of our most important but least obvious natural resources. Several countries rely on groundwater pumped to the earth's surface for drinking water. Some people in rural locations rely solely on it for fresh water. Toxins from landfills and septic systems infiltrate an aquifer and poison groundwater, rendering it unfit for human consumption.

Nitrates from the overuse of chemical fertilizers or manure on farmland are also common groundwater pollutants, as are petroleum products leaking from underground storage tanks and diesel fuel from gas stations, leaching of fluids from landfills and dumpsites, road salt, accidental spills, and excessive applications of chemical pesticides, as shown in Figure 4.5.

Pollutant removal from groundwater can be complex and costly. Poisoned aquifers may be uninhabitable for decades, if not thousands of years. Groundwater can transmit toxins far from the initial contaminating source as it seeps into streams, lakes, and seas.

To prevent groundwater pollution, the best way is to properly dispose of all waste and avoid dumping chemicals down drains or on the ground, to continuously test underground fuel oil tanks for leaks, and, if possible, to replace them above ground, to safely store all chemicals and fuels, to minimize the use of chemicals, and to always use them according to directions.

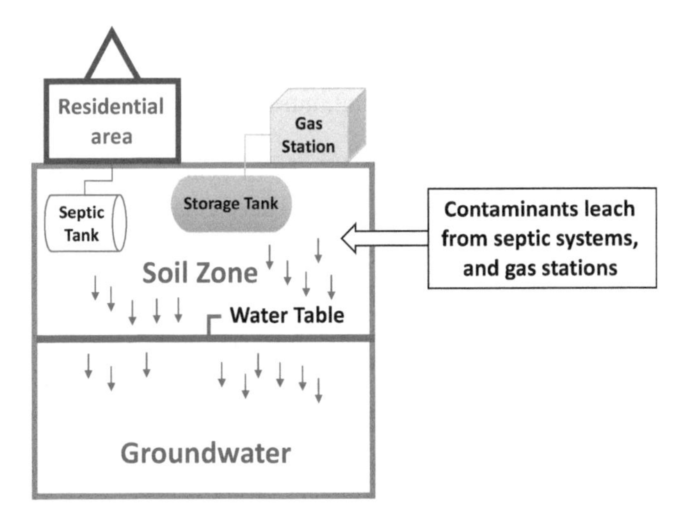

Figure 4.5: Groundwater pollution.

4.3.2 Surface Water Pollution

Many of the world's rivers, streams, and lakes are poisoned, making swimming, fishing, and drinking unhealthy. Nutrient pollution, which includes nitrates and phosphates, is the most common type of contamination in these freshwater sources. While these nutrients are necessary for plant and animal life, agricultural waste and fertilizer runoff have made them major pollutants. The municipal and industrial waste also releases toxins into the environment. There's also the litter that corporations and individuals toss directly into waterways.

From farms, factories, and towns, streams and rivers also transport pollutants like pesticides, fertilizers, and heavy metals into our bays and estuaries, where they eventually make their way to the seas and oceans. While this is going on, maritime debris, particularly plastic, is being blown in by the wind or washing up through storm drains and sewers. Along with continuously absorbing carbon pollution from the atmosphere, seas and oceans are occasionally polluted by large and small oil spills and leaks.

4.4 Effects of Water Pollution

4.4.1 On Human Health

Waterborne pathogens are bacteria and viruses that cause disease and are present in human and animal waste, and they are one of the most common reasons of illness caused by contaminated drinking water. Among the diseases spread by contaminated water are cholera, giardiasis, and typhoid. Even in affluent countries, sewage treatment plant emissions, as well as farm and urban runoff, introduce potentially harmful bacteria into rivers.

Another crisis is coming from lead poisoning, caused by cost-cutting initiatives and deteriorating water infrastructure, which provides a sharp reminder of how harmful chemical and other industrial pollutants in our water may be. The problem extends far beyond flint and encompasses far more than lead, as various chemical pollutants, ranging from heavy metals like arsenic and mercury to pesticides and nitrate fertilizers, are finding their way into our water systems. If taken, toxins can cause various health problems ranging from cancer to hormone disruption to impaired brain function.

4.4.2 On the Environment

An intricate web of interactions between organisms such as plants, animals, bacteria, and fungi is necessary for the maintenance of healthy ecosystems. Any harm to one of these creatures has the potential to set off a cascade that puts entire aquatic ecosystems in danger. When a lake or sea has an algal bloom due to water pollution, the abundance of recently provided nutrients drives plant and algae growth, lowering water oxygen levels. Eutrophication, or a lack of oxygen, causes plants and animals to suffocate, which can result in “dead zones,” or regions of water that are virtually lifeless. Rarely, these harmful algal blooms can produce neurotoxins that are harmful to animals like whales and sea turtles.

Chemicals and heavy metals from municipal and industrial effluent that contaminate streams present another issue. These pollutants, which are poisonous to aquatic life and frequently shorten an organism’s life expectancy and capacity for reproduction, move up the food chain as predators devour their prey. This is how large fish like tuna and others pick up large amounts of poisons like mercury.

Marine debris, which can starve, choke, and strangle creatures, is also a threat to marine ecosystems. The majority of this solid waste, including plastic bags and drink cans, is flushed into storm drains and sewers before being discharged at sea, transforming our oceans into a soup of junk that occasionally accumulates to form floating garbage patches. Many marine life species have been wiped out by discarded fishing gear and other sorts of rubbish.

Another issue is ocean acidification, which makes it difficult for shellfish and coral to survive. Oceans are growing more acidic despite the fact that they absorb around a quarter of the carbon pollution produced each year by the combustion of fossil fuels. This process makes shellfish and other species' shell formation more difficult, and it may have an impact on the neurological systems of sharks, clownfish, and other marine animals.

4.5 Water Pollutants

Domestic trash, insecticides and herbicides, food processing waste, livestock operations waste, volatile organic compounds, heavy metals, and chemical waste are all examples of water pollutants. Some of these contaminants are summarized in Table 4.1.

Table 4.1: Examples of water pollutants.

Fertilizers	Pesticides
Nitrates	Phosphates
Dissolved inorganic ions	Dissolved organic compounds
Plastics	Perchloroethylene (PERC)
Dissolved gases	Polyfluoroalkyl substances (PFASs)
Bacteria	Parasites
Viruses	Polycyclic aromatic hydrocarbons (PAHs)
Suspended particles	Colloidal particles
Lead	Mercury
Cadmium	Arsenic

4.6 Questions

4.6.1 List the five consequences of water pollution.
4.6.2 What are the known causes of water pollution?
4.6.3 Does agriculture contribute significantly to environmental pollution and how?
4.6.4 What are the most common agricultural pollutants?
4.6.5 What does the term sewage stand for?
4.6.6 Give examples of the sewage spill's effects.
4.6.7 Name five techniques used to clean up oil spills.
4.6.8 What are the common groundwater pollutants?
4.6.9 Can the groundwater pollutants be removed?
4.6.10 Give five examples of water pollutants.

Chapter 5
Microbial Contaminants in Water

Water is necessary for survival; however, many people lack access to clean and safe drinking water, and many people die as a result of waterborne bacterial illnesses. The most common bacterial diseases transmitted through water are cholera, typhoid fever, and bacillary dysentery, and the most well-known contaminants of concern in drinking water are bacteria like *Helicobacter pylori*, the *Salmonella* family, *Escherichia coli*, and viruses like hepatitis A, Norwalk-type viruses, rotaviruses, adenoviruses, enteroviruses, and reoviruses. Figure 5.1 depicts the four most typical sources of bacterial contamination.

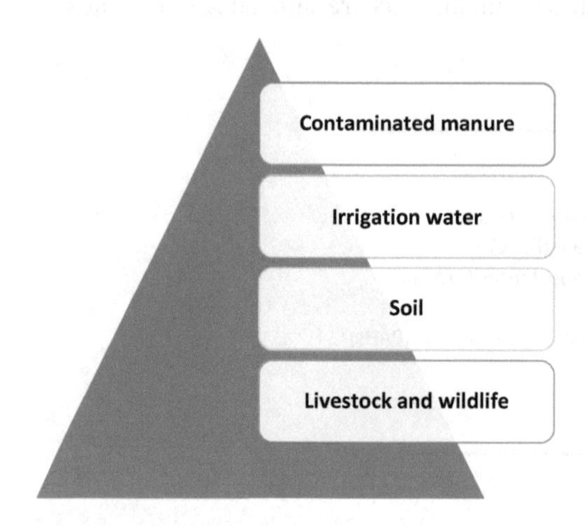

Figure 5.1: Bacterial contamination sources.

It is not always easy to identify if water has microbiological contamination, but there are some indications that can help. A summary of them is shown in Figure 5.2.

One of the biggest issues is providing everyone with access to safe drinking water, and drinking water should always be microbiologically controlled. Drinking water should undergo a routine basic microbiological investigation to check for the presence of several bacteria that commonly cause bacterial illnesses.

5.1 Water Consumption as a Source of Disease

Life is dependent on water. Everyone should have access to a sufficient, secure, and reliable supply. Enhancing access to clean drinking water can have a positive impact on health. To ensure the highest level of safety for drinking water, every effort should

https://doi.org/10.1515/9783111332468-005

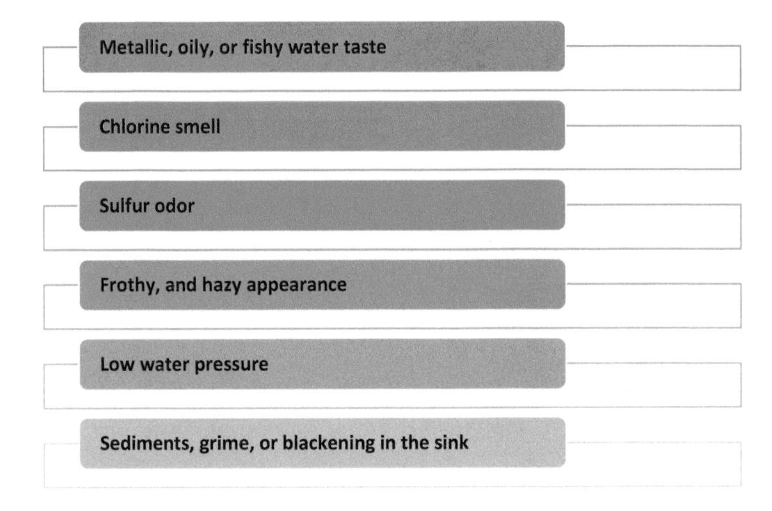

Figure 5.2: Signs for water microbiological contamination.

be made. Many individuals have trouble getting access to clean water. In industrialized nations, a clean and treated water supply to every home may be the standard, but in developing nations, access to clean water and sanitation is not the norm, and waterborne diseases are frequent.

Millions of people die each year from diseases linked to water, says the WHO. More than 50% of them are microbial intestinal diseases, with cholera being the most notable. Generally speaking, drinking water that has been tainted with human or animal feces poses the biggest microbiological dangers. Pathogenic fecal bacteria, including those found in fresh water and coastal seawaters, are primarily obtained via wastewater discharges.

The following paragraphs discuss the most common bacterial infections transmitted through water, including cholera, salmonellosis, and shigellosis or bacillary dysentery, as well as dangerous *E. coli* strains.

5.1.1 Cholera

Cholera is caused by curved, Gram-negative bacillus called *Vibrio cholerae*. The species is a naturally occurring, free-living organism that can be found in a variety of aquatic environments. Where there is poor water treatment, sanitation, and hygiene, cholera is more likely to occur and spread. The severe diarrheal disease cholera is brought on by the bacterial pathogen *V. cholerae*. Figure 5.3 shows the four cholera symptoms.

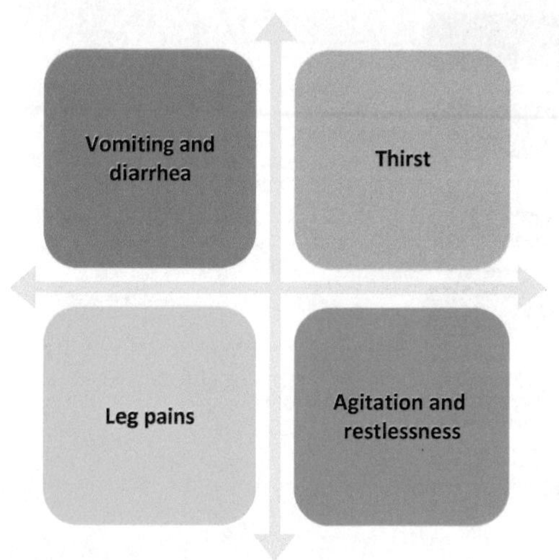

Figure 5.3: Cholera symptoms.

Cholera infection treatment involves the use of doxycycline as a first-line treatment. If doxycycline resistance is documented, and azithromycin and ciprofloxacin are viable alternatives.

5.1.2 Salmonellosis

Salmonellosis is a frequent bacterial illness of the intestine caused by *Salmonella* infection. *Salmonella* bacteria normally dwell in the intestines of animals and humans and are excreted in the form of stool.

Humans are most commonly infected through polluted water or contaminated food. Figure 5.4 summarizes the possible signs and symptoms of salmonella infection.

Salmonella infections are typically treated with fluoroquinolones or third-generation cephalosporins, such as ciprofloxacin and ceftriaxone.

5.1.3 Shigellosis or Bacillary Dysentery

Shigellosis is a digestive infection caused by *Shigella*, a Gram-negative, straight rod-like bacteria that does not produce spores. *Shigella* is quickly transmitted from person to person, and just a little amount of *Shigella* is required to induce sickness. *Shigella* bacteria cause infection when consumed by mistake from polluted water. *Shigella* germs can enter water through sewage or through a *Shigella*-infected swimmer. The signs and symptoms of *Shigella* infection are summarized in Figure 5.5.

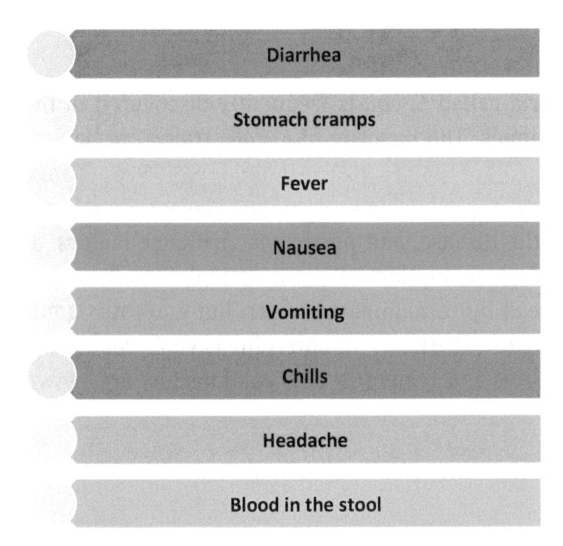

Figure 5.4: Signs and symptoms of salmonella infection.

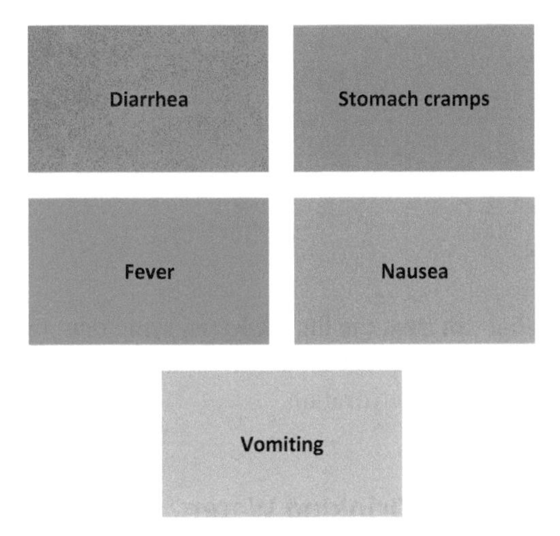

Figure 5.5: Signs and symptoms of *Shigella* infection.

Combating dehydration by drinking plenty of fluids, especially electrolyte solutions, many of which are available over the counter, is the first line of *Shigella* infection treatment. Infections that are moderate to severe may necessitate medical attention. Antibiotics are typically used to clear the germs from the digestive tract. Azithromycin (Zithromax), ciprofloxacin (Cipro), and sulfamethoxazole/trimethoprim (Bactrim) are examples of strong antibiotics.

5.1.4 Pathogenic *Escherichia coli* Strains

A Gram-negative, rod-shaped bacteria called *E. coli* is frequently discovered in the lower intestine of warm-blooded animals. The majority of *E. coli* strains are benign, but pathogenic types can seriously affect humans by leading to septic shock, meningitis, or urinary tract infections. Based on epidemiological data, morphological characteristics, clinical characteristics of the disease, and particular virulence factors, at least six primary groupings of *E. coli* strains isolated from intestinal disorders have been identified. They can all be spread by contaminated water, but **e**ntero**t**oxigenic ***E. coli*** **(ETEC)**, **e**ntero**h**emorrhagic ***E. coli*** **(EHEC)**, and **e**ntero**i**nvasive ***E. coli*** **(EIEC)** are of exceptional relevance. A few signs and symptoms of E. coli infection are shown in Figure 5.6.

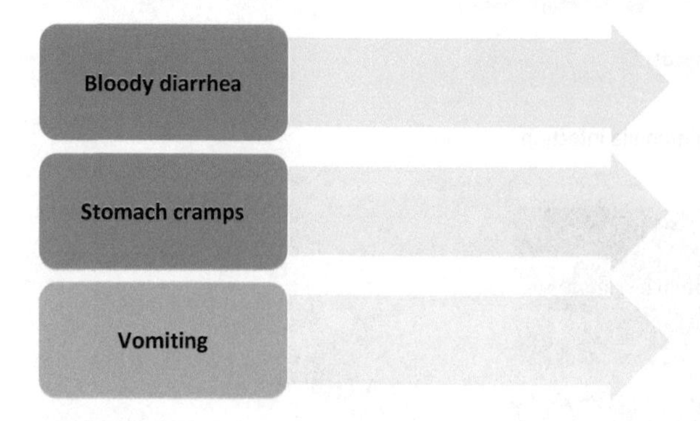

Figure 5.6: Signs and symptoms of *E. coli* infection.

There are no drugs available right now that can treat the illness, lessen symptoms, or prevent issues brought on by *E. coli*. The most frequent kind of treatment for most people is rest. Fluids help to reduce tiredness and dehydration.

5.2 Microbial Contaminants in Potable Drinking Water

The objective of water treatment facilities is to create drinking water that is free of pathogens and parasites, not necessarily sterile water. However, a potable water supply can become contaminated from a number of different sources, including:
- Water supply source
- Water treatment facility
- Water distribution system
- Biofilms formation

5.3 Bottled Water Microbiology

Numerous parasites and pathogenic microbes, as well as native microorganisms, are frequently present in water sources and could end up in drinking water. Bacterial pathogens, viral pathogens, and protozoan parasites are the three types of pathogens most frequently seen.

Even in established industrial nations with state-of-the-art water purification facilities, effective marketing and advertising have contributed to the expansion of the bottled water industry globally.

For a variety of reasons, including complaints about the taste of tap water and worries about the presence of hazardous chemicals, germs, and parasites, people choose to drink bottled water. Consumers believe bottled water to be pure and harmless; however, multiple studies have found that it can occasionally include hazardous chemicals and pathogenic microbes that could have negative effects on human health.

The main purpose of bottled water today is for drinking, but it has also been used to make baby formula, clean contact lenses, and run humidifiers. Customers typically believe bottled water to be of higher quality and greater flavor than tap water. However, it's not always true that bottled water is of greater quality.

According to some investigators, the quality of bottled water isn't any higher than that of municipal drinking water. Additionally, it costs a hundred times as much as regular tap water. Additionally, plastic bottle collecting, packing, transportation, and disposal harm the environment. It's possible that source water or water treatment prior to bottling is to blame for the seldom presence of harmful compounds and pathogenic microbes.

5.3.1 Source Water

Aquatic native microbes, possible pathogens, and parasites may come from the source water or may be added at the bottling facility through contaminated equipment. The identification of indigenous microorganisms in natural mineral fluids and spring waters has been aided by culture-dependent and molecular-based approaches.

5.3.2 Water Treatment Before Bottling

Depending on the source, activated carbon, sand filtration, microfiltration, distillation, deionization, softening, mineral adjustment, UV radiation, carbonation, ozonation, or reverse osmosis are few of the methods that can be used to treat water before bottling.

5.4 Questions

5.4.1 What are the typical bacterial contamination sources?
5.4.2 Give examples of the most prevalent bacterial infections that are spread by water.
5.4.3 List three signs of microbial contamination.
5.4.4 What are the four signs of cholera disease?
5.4.5 How is the infection caused by cholera treated?
5.4.6 Identify the *Salmonellosis*-causing bacterium.
5.4.7 What symptoms and signs indicate a *Shigella* infection?
5.4.8 What does ETEC stand for?
5.4.9 What are the sources of water supply contamination?
5.4.10 Does bottled water always have high quality?

Chapter 6
Water Analysis

Water testing and analysis can look at a variety of characteristics and parameters, including those that are physical such as temperature, color, and turbidity, chemical such as the presence of metals and compounds, and microbiological such as contamination with bacteria and viruses. To assess the quality of the water, chemists, and technicians measure a variety of parameters. These include the following: temperature, pH, dissolved solids, particle matter, dissolved oxygen, hardness, and suspended sediment.

Water testing on a regular basis can assist in ensuring that water supplies fulfill appropriate health and safety standards, safeguarding customers from potential harm. The laboratory is where exact work is done to identify proper analysis and testing, as well as to monitor the quality of the produced water. Representative water samples aid in determining the purity and quality of water sources.

Water sampling is critical for producing accurate water analysis results. The right technique, as well as glasses or plasticware, must be used to get the desired result. Metals, for example, react with glass and can adsorb on it; thus, water samples for metal analysis are collected in polyethylene containers. Another important factor to consider is the appropriate lifetime of the parameters to be employed as well as the preservation technique.

Separating the dissolved or suspended components in the sample before analysis is a common requirement in water analysis. This is usually accomplished by a physical separation stage. The most prevalent form of sample pretreatment is filtration. Membrane and glass fiber filters come in various pore diameters, while membranes come in a variety of materials. In general, various tools are used for water analysis, some of which are listed in Figure 6.1.

The theoretical underpinnings of the typical water tests used for water analysis are covered in detail in the pages that follow.

6.1 Common Water Tests

6.1.1 Color, Odor, and Taste

Natural inorganic and organic chemical pollutants, biological sources or processes, manmade chemical pollution, corrosion, or problems with water treatment methods like chlorination can all contribute to color, taste, and odor (smell). Drinking water in particular may have an unpleasant taste or odor as a result of pollution or a problem with the water treatment or distribution system. Table 6.1 summarizes water color, odor (smell), taste, and related pollutants.

https://doi.org/10.1515/9783111332468-006

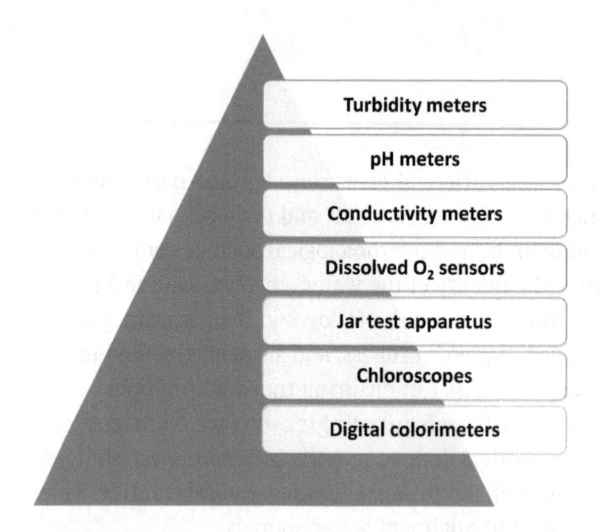

Figure 6.1: Water analysis tools.

Table 6.1: Water color, odor (smell), taste, and related pollutants.

Water color	Possible pollutants
Bluish-green or green	High concentrations of copper, lead, zinc, or other trace metals
Grayish	High levels of aluminum
Blackish	High quantities of copper, iron, and manganese sulfides or oxides as well as sulfate-related, iron, or manganese nuisance bacteria
Tea-like appearance	Elevated tannic and fulvic acid levels
Water odor or smell	**Possible pollutants**
Phenolic strong odor	Petrochemicals and volatile organic compounds
Chemical odor	Synthesized organic molecules and industrial chemicals
Methane-like odor	Mercaptans
Oily odor	Gasoline, petrochemicals, and surfactants
Smell of rotten eggs	Sulfur, hydrogen sulfide, and microbial pollutants
Fishy odor	Surfactants, barium, cadmium, and algal blooms in surface water sources
Musty or earthy odor	Iron bacteria, mold, fungi, algal blooms, or a high bacterial count
Cucumber smell	Iron bacteria and algal blooms
Chlorine smell	High quantities of chlorine and chlorine by-products

Table 6.1 (continued)

Water taste	Possible pollutants
Metallic	High levels of metals like iron, manganese, copper, lead, and zinc
Oily, fishy, or perfume-like, hazy, and bitter	Breakdown of waste products

6.1.2 Turbidity

Turbidity is a measure of how much transparency the water loses due to the presence of suspended particulates. It can also be defined as a fluid's cloudiness or haziness induced by a large number of individual particles.

A nephelometer, often known as a turbidity meter, is the best tool for measuring turbidity in a wide range of samples. An electronic handheld meter that measures light scatter using a light and photodetector and outputs turbidity units such as **n**ephelometric **t**urbidity **u**nits **(NTUs)** or **f**ormazin **t**urbidity **u**nits **(FTUs)**. The nephelometric approach compares the quantity of light scattered in a water sample to the amount of light scattered in a control solution. There are various types of turbidity meters:

- Benchtop meters are used to take measurements on samples that can be carried to a laboratory.
- Submersible meters can be submerged in water and used to measure turbidity on-site.
- Continuous flow meters provide continuous turbidity monitoring as a stream of water passes past the sensor.

Drinking water turbidity should preferably be kept below 1 NTU. The lower the turbidity, the less particle matter there is. Low turbidity readings imply clear water, whereas high values suggest cloudy water. In large, well-managed municipal supplies, less than 0.5 NTU should be maintained at all times with an average of 0.2 NTU or less, regardless of source water type and quality.

When it comes to turbidity levels, less than 10 NTUs is considered low turbidity, more than 50 NTUs is considered moderate, and more than 100 NTUs is considered severe. The World Health Organization recommends that the turbidity of drinking water be less than 1 NTU and no higher than 5 NTUs.

High turbidity in drinking water is both unpleasant and potentially harmful to one's health. Pathogens can find food and refuge in turbidity. If the reasons for high turbidity are not eliminated, they can stimulate the regeneration of bacteria in the water, resulting in waterborne disease outbreaks. Figure 6.2 summarizes the elements that contribute to water turbidity.

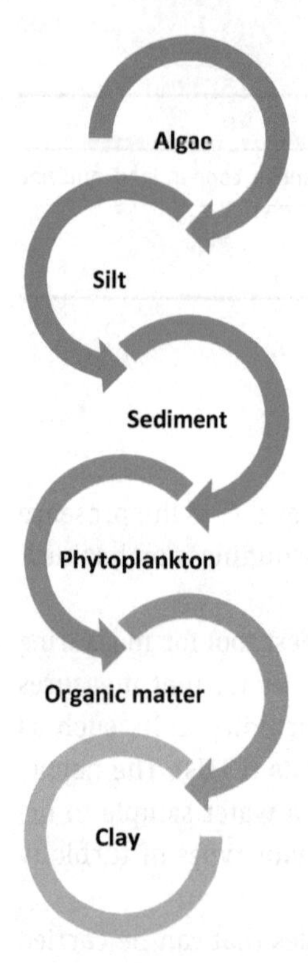

Figure 6.2: Factors that contribute to water turbidity.

Visual tools, in addition to the turbidity meters discussed above, can be used to measure turbidity in water. Turbidity can be measured visually using two techniques that are commonly used in lake and stream monitoring programs:

- Secchi disks are black and white circular disks that are used to determine the clarity of water. Secchi disks are an inexpensive, portable, and simple technique to obtain turbidity readings in a big body of water. The accuracy of the results is determined by the lighting and the person's eyesight.
- Transparency tubes are transparent tubes with a Secchi disk and a release valve at the bottom. The tubes also have measurement markings on them to measure the volume of water collected.

The following precautions should be observed during a turbidity test and when using a nephelometric turbidimeter:

- Clean glass or plastic containers should be used to collect samples.
- Samples should be analyzed as soon as possible after collection.
- Dirty, scarred, or chipped sample tubes might give high results.

6.1.3 Jar Test

Coagulation–flocculation is the process of aggregating tiny particles in water into larger, heavier clumps that settle out quickly. The larger particles are referred to as flocs. Properly constructed floc will swiftly settle out of the water in the sedimentation basin, removing the majority of the turbidity. Many plants require the operator to modify coagulant dosages at regular intervals to achieve optimal coagulation due to changing water properties. A jar test is performed to compare various coagulant dosages to estimate the minimum coagulant dose required to achieve certain water quality goals. Figure 6.3 depicts a schematic diagram of the jar test apparatus.

This test is generally accomplished by replicating the treatment plant settings for each jar test and pouring 1 L of representative water sample into each jar, followed by a sequence of doses of the required chemical, beginning with control (0 dose) and ending with the highest dose; then, using the mixing paddles to mix the chemicals and samples before treating them in plant conditions and checking the results.

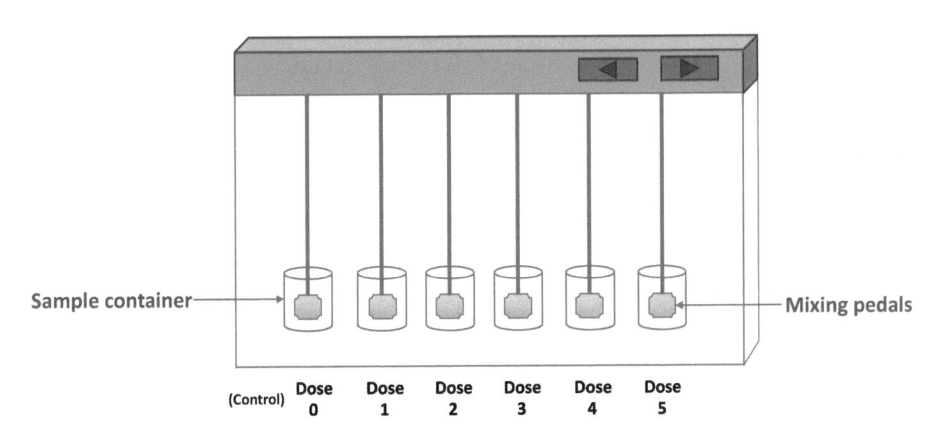

Figure 6.3: A schematic diagram of jar test apparatus.

6.1.4 pH Measurements

The concentration of hydrogen ions is determined by pH. The hydrogen ion concentration of a solution, measured in moles per liter, is expressed as the negative logarithm of pH. For instance, pure water has a pH of 7 and a hydrogen ion concentration of 10^{-7} mol/L at ambient temperature. Normally, the pH scale ranges from 0–14. At a pH of 0, acids frequently fully ionize in water. A solution that has a pH value less than 7 indicates a higher concentration of hydrogen ions and is regarded as acidic. Water has a pH of 14.0 when a fully ionized base or alkaline solution is present.

pH values are measured and controlled with pH meters; a schematic diagram of the pH meter is shown in Figure 6.4. The pH meter's probe is placed into the water sample. At the tip of the probe, there is a thin glass bulb that contains a reference electrode, usually made of silver/silver chloride (Ag/AgCl). The glass bulb itself serves as the pH-sensitive glass electrode.

In order to obtain an accurate measurement, pH meters must be calibrated. To accomplish this, the probe is taken out of the solution and cleaned with deionized water before being put into a buffer with a specified pH for calibration. After being taken out of the buffer and cleaned with deionized water, the electrode is then reinserted into the water solution to be measured.

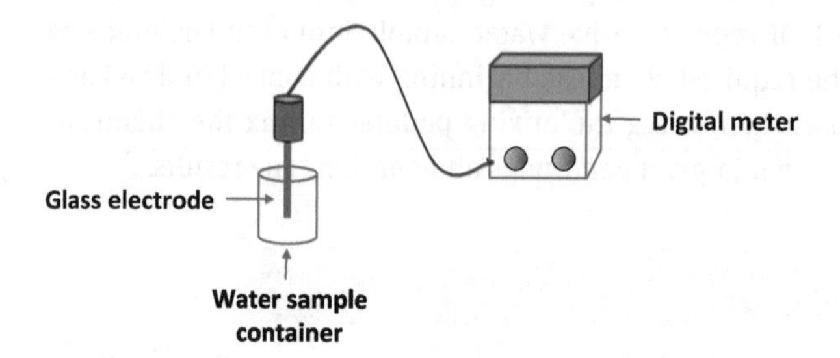

Figure 6.4: A schematic diagram of the pH meter.

Many water testing and purification techniques rely on pH monitoring. For the following reasons, water pH should be tested and monitored:
- A change in water pH can affect the behavior of compounds in the water.
- pH has an impact on product quality and customer safety.
- pH changes can impact flavor, color, shelf life, product stability, and acidity.
- Tap water with an improper pH can cause corrosion in the distribution system and allow dangerous heavy metals to leach out.
- Managing the pH of industrial water settings aids in the prevention of corrosion and equipment damage.
- In natural ecosystems, pH can have an impact on plants and animals.

6.1.5 Alkalinity

Alkalinity is a measurement of the amount of dissolved alkaline compounds in water that may neutralize acid (pH greater than 7.0). The three major types are as follows:
- Bicarbonate (HCO_3^-)
- Carbonate (CO_3^{2-})
- Hydroxides

The general classification system for alkalinity is as follows:

- Low alkalinity < 20 mg $CaCO_3$/L
- Moderate alkalinity 20–160 mg $CaCO_3$/L
- High alkalinity > 160 mg $CaCO_3$/L

Alkalinity can occur in water in three basic forms depending on the pH: carbonate, bicarbonate, or hydroxide. The sum of these three kinds is total alkalinity. Alkalinity is determined by titrating a water sample with standard acid to a specific pH and recording the result as P, M, or T alkalinity. P alkalinity is titrated with phenolphthalein to pH 8.3, M alkalinity is titrated to pH 4.6 using the methyl orange indicator, and T alkalinity is titrated to pH 4.5 using the total alkalinity indicator. As can be seen, the endpoints of both the methyl orange indicator (pH 4.6) and the total alkalinity indicator (pH 4.5) are nearly identical, so they can be used interchangeably for total alkalinity readings. As a result, the values of M and T can also be interchanged. Once P, M, or T alkalinity values are determined, Table 6.2 can be use to calculate treatment control and effectiveness.

Table 6.2: Alkalinity relationship.

Titration results of alkalinity test	Hydroxide alkalinity (OH^-)	Carbonate alkalinity (CO_3^{2-})	Bicarbonate alkalinity (HCO_3^-)
P = 0	0	0	T
P < ½ T	0	2P	T–2P
P = ½ T	0	T	0
P > ½ T	2P–T	2(T–P)	0
P = T	T	0	0

Note that values of M and T can be interchanged to P/T or P/M.

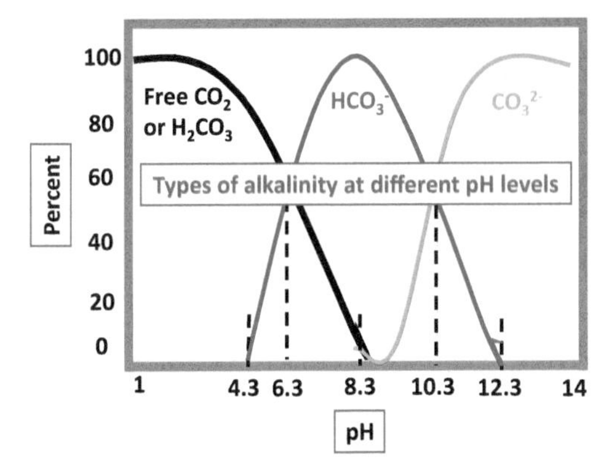

Figure 6.5: Types of alkalinity at different pH levels.

As depicted in Figure 6.5, alkalinity in water starts to occur with a pH of 4.3. Anything with a pH below 4.3 has only dissolved CO_2 (or carbonic acid) and no alkalinity. In essence, bicarbonate alkalinity (blue line) starts to change from the carbonic acid (black line) at pH 4.3. The change from bicarbonate to carbonate ions then takes place at a pH of 8.3 (green line). More hydrogen dissociates and evaporates at higher pH levels. As a result, bicarbonate (HCO_3^-) is created when carbonic acid (H_2CO_3) loses hydrogen.

Bicarbonate loses hydrogen when the pH rises; at a pH of 8.3 and becomes carbonate (CO_3^{2-}). It is evident that the bicarbonate–carbonate equilibrium predominates at pH values between 8.3 and 12.3, with the CO_2–bicarbonate equilibrium existing in the pH range of 4.3–8.3. Additionally, the pH–carbonate equilibrium shows:

- Only dissolved CO_2 (gas) or carbonic acid is present at a pH of 4.3.
- Only carbonic acid and bicarbonate ions exist between pH 4.3 and 8.3, with the amount of each species varying depending on the pH.
- Only bicarbonate ions are present at pH 8.3.
- Only bicarbonate and carbonate ions are present between pH 8.3 and 12.3, with the amount of each species varying depending on the pH.
- At pH 6.3 and pH 10.3, buffers are formed.

6.1.6 Hardness

The amount of measured divalent metal cations is referred to as water hardness. Only two divalent cations, dissolved calcium (Ca^{2+}) and dissolved magnesium (Mg^{2+}) are present in significant amounts in most waters. Both calcium and magnesium are mainly bound to bicarbonate, sulfate, or chloride in natural water. Bicarbonate changes to carbonate and precipitates out with Ca^{2+} to produce calcium carbonate ($CaCO_3$) scale when hard water evaporates or is heated above 61 °C. Since $CaCO_3$ is not a component of water, water hardness levels are normally expressed in mg/L as $CaCO_3$ equivalent. There are various classification systems for indicating the degree of hardness, but generally speaking, soft water has $CaCO_3$ equivalents under 60 mg/L and hard water has more than 120 mg/L. According to general recommendations, water should be categorized as soft up to 60 mg/L of calcium carbonate, moderately hard between 61 and 120 mg/L, hard between 121 and 180 mg/L, and very hard over 180 mg/L.

Water hardness is of two types: carbonates and noncarbonate hardness. The carbonate and bicarbonate salts of calcium and magnesium are the main contributors to carbonate hardness (temporary hardness). Magnesium and calcium hydrogen carbonates ($Ca(HCO_3)_2$ and $Mg(HCO_3)_2$) are responsible for their occurrence.

When heated, both species break down, and $CaCO_3$ or $MgCO_3$ precipitates. As a result, boiling water can be used to eliminate temporary hardness.

The sulfates and chlorides of magnesium and calcium cause noncarbonate hardness, also known as permanent hardness. In other words, water that contains calcium sulfate ($CaSO_4$), calcium chloride ($CaCl_2$), magnesium sulfate ($MgSO_4$), or magnesium chloride ($MgCl_2$) becomes permanently hard.

Total hardness is expressed as the sum of carbonate hardness. Water hardness is expressed in parts per million (ppm) and milligrams per liter (mg/L):

$$\underset{(\text{mg/L as } CaCO_3)}{\text{Total hardness}} = \underset{(\text{mg/L as } CaCO_3)}{\text{calcium hardness}} + \underset{(\text{mg/L as } CaCO_3)}{\text{magnesium hardness}}$$

$$= 2.50 \times \text{calcium conc.}\left(\text{mg/L as } Ca^{2+}\right) + 4.12 \times \text{magnesium conc.}\left(\text{mg/L as } Mg^{2+}\right)$$

or

$$\text{Hardness} = 2.497\,\text{Ca} + 4.118\,\text{Mg}$$

Total hardness, calcium hardness, and magnesium hardness are the three parts used to analyze water hardness. An ammonia buffer solution is added to the sample in order to keep the pH around 11 while measuring total hardness. The next step is to add sodium 1-(1-hydroxy 2-naphthylazo)-6-nitro-2-naphthol-4-sutphonate, **E**riochrome **B**lack **T** (**EBT**; Figure 6.6A) indicator to allow the sample's wine-red color to fully develop. The sample is then titrated against **e**thylene**di**amine**t**etra**a**cetic acid (**EDTA**) solution (Figure 6.6B) after that and the sample's color will change from wine red to blue. The titration of total hardness comes to an end at this stage. Ammonium purpurate, murexide (Figure 6.6C) indicator is added to the sample, after the addition of sodium hydroxide solution to the sample to maintain pH, to estimate the calcium hardness by titration. When the sample is titrated against an EDTA solution, the sample's color will change from pink to purple. The calcium hardness test comes to an end at this point.

Total hardness minus calcium hardness value equals magnesium hardness:

$$\text{Magnesium hardness} = \text{total hardness} - \text{calcium hardness}$$

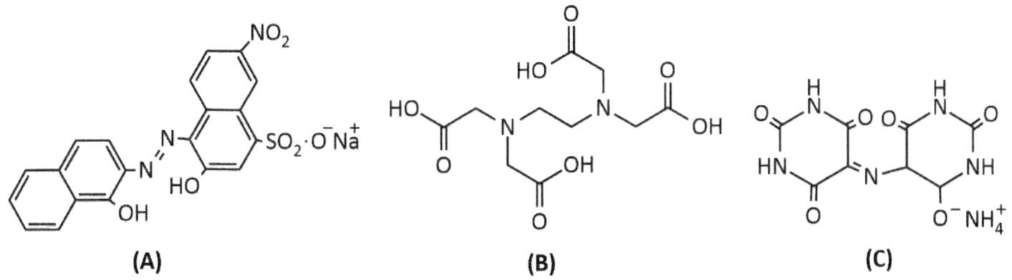

Figure 6.6: Chemical structures of EBT, EDTA, and murexide.

6.1.7 Residual Chlorine

Residual chlorine is the quantity of chlorine that remains in water after a specific period of time or contact time. The residual chlorine test is one of the most typical tests performed in water treatment facilities. The residual chlorine test detects the quantity of residual chlorine in the water that has passed testing and is appropriate for distribution.

The fact that drinking water still contains chlorine after initial chlorine addition to inactivate bacteria and some viruses shows that the water was properly stored to prevent recontamination. In water distribution systems, free chlorine residuals of 0.5 to 2.0 mg/L offer efficient continuous disinfection.

The WHO has set a limit of 5 mg/L for the residual amount of free chlorine in drinking water. The WHO recommends 0.2 mg/L as the lowest acceptable level of free chlorine in treated drinking water. There are three forms of residual chlorine in water treatment:

- Free: composed of dissolved hypochlorite ions, hypochlorous acid, and chlorine gas.
- Combined: composed of chloramines that can kill bacteria and oxidize organic matter.
- Total: the sum of free and combined residual chlorine.

The amount of residual chlorine left in the chlorinated water after the required contact period can be experimentally determined by using any of the following tests:

- Diethyl-*p*-phenylene-diamine test
- Orthotolidine test
- Chloroscope test
- Digital colorimeters

6.1.7.1 Diethyl-*p*-Phenylenediamine Test

Chlorine and chlorine-released compounds are commonly used in drinking water disinfection, for microbial growth control in cooling water, and in many other water treatment systems. Accurate detection of the chlorine residual is critical for controlling these chlorination systems. Although chlorine levels can be represented in terms of free chlorine, mixed chlorine, or total chlorine residuals, the free chlorine residual is the most essential for the majority of applications.

A pink color results from the reaction of free chlorine with *N,N*-**d**iethyl-*p*-**p**henylene **d**iamine **(DPD**; Figure 6.7), in a buffered solution. The relationship between color intensity and free chlorine content is linear. After the excess potassium iodide is added, any combined chlorine that is present undergoes a reaction. The increase in intensity now indicates the combined chlorine concentration and the color intensity is now inversely proportional to the total chlorine concentration. Free and combined chlorine in the sample can be differentiated using this method.

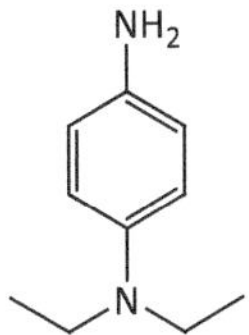

Figure 6.7: Chemical structure of *N,N*-diethyl-*p*-phenylenediamine.

DPD test is not only used to determine the quantity of chlorine in water but also bromine, iodine, and other disinfectants. It can distinguish between free and combined chlorine. The color is tested on a comparator so that it may be easily compared to the reference colors. There are two chlorine concentration levels available for the comparator disks: 0.1–1.0 mg/L and 0.2–4.0 mg/L.

To do the test, 10 mL of water is collected in the comparator tube to begin the test; then a reagent tablet (DPD chlorine tablet) is added, and water is shaken to dissolve the tablet completely. Compare the resultant color with reference colors to get the residual chlorine level in mg/L.

When measuring chlorine, there are four common DPD tablets that can be used:

- DPD 1 measures free chlorine.
- DPD 2 measures combined chlorine when used with DPD 1.
- DPD 3 measures total chlorine when used with DPD 1.
- DPD 4 calculates the total chlorine available or total chlorine content.

6.1.7.2 Orthotolidine Test

Orthotolidine, (Figure 6.8A), not orthotoluidine (Figure 6.8B), as some researchers have misidentified it, is used in this test. In the figures, both chemical structures are depicted. To conduct this test, 10 mL of chlorinated water is collected in a glass tube after the required contact period. This is mixed with 0.1 mL of orthotolidine solution. After 5 min, the color created is observed. The presence of total chlorine in the water is generally indicated by the production of yellow color. The higher the chlorine residual, the more yellow the color. By comparing the color formed in the glass tube to the standard colors already preserved in the laboratory, the amount of residual chlorine can be determined.

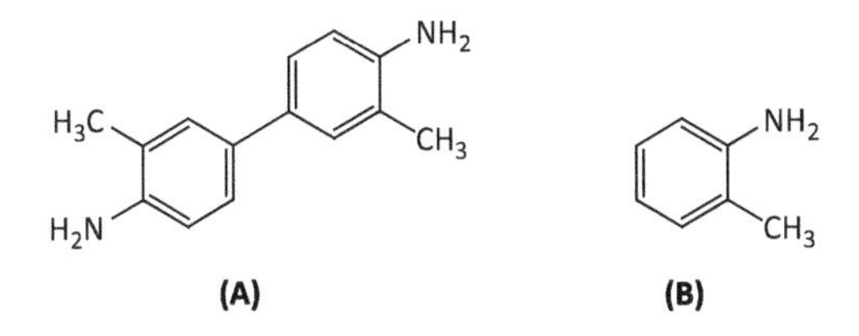

Figure 6.8: Orthotolidine versus orthotoluidine chemical structures.

6.1.7.3 Chloroscope Test

A chloroscope is used as a comparison instrument in this test. The viewing window on the right is used for the water sample, which is then compared with reference colors on the left side on the comparator disk. The chloroscope detects accessible chlorine levels ranging from 0.1–2.0 ppm (mg/L). If chlorine is present, an orthotolidine solution causes the water to turn yellow or greenish-yellow. To perform this test, fill a test tube to 5 mL with the water sample to be tested. Using a dropper, add 1–2 drops of *O*-tolidine solution to the sample. Shake well and leave the tube for a few minutes to allow the color to develop.

As soon as the color is generated, insert the tube into the hole on the right side of the comparator to compare it to standard colors and record the reading from the ppm (mg/L) scale marked in front of the chloroscope, as illustrated in Figure 6.9.

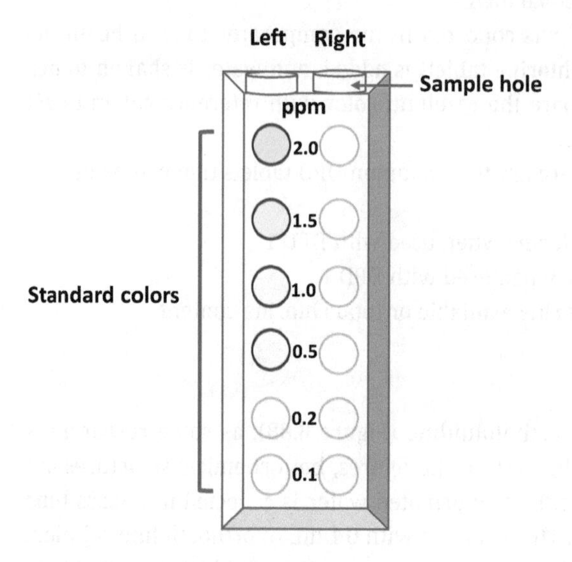

Figure 6.9: A chloroscope diagram.

6.1.7.4 Digital Colorimeters

With digital colorimeter or pocket colorimeter (Figure 6.10), the free chlorine and/or total chlorine residual measurements are the most precise. The color of the sample water turns pink when DPD tablets or powder are added, and the vial is then put in the colorimeter, which reads the intensity of the color change by emitting a wavelength of light and digitally calculating and displaying the color intensity (free and/or total chlorine residual). The colorimeter records values between 0 and 4 mg/L (ppm).

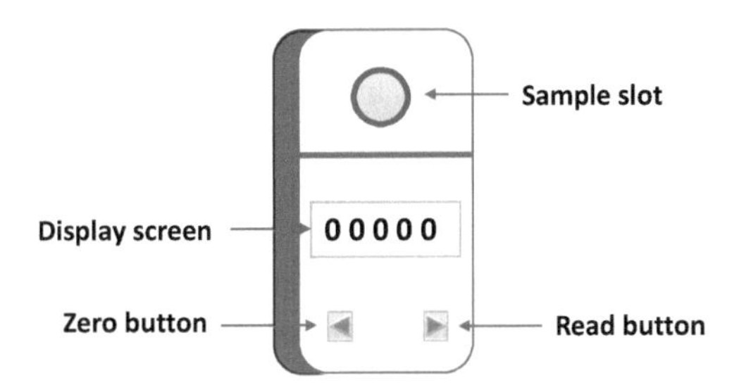

Figure 6.10: Pocket colorimeter diagram.

6.1.8 Chlorides

In most natural waters, chloride is a naturally occurring element that is frequently present as a component of salt (sodium chloride) or, in certain situations, in combination with potassium or calcium. Water can taste bitter at chloride concentrations above 250 mg/L; however, the threshold is dependent on the related cations.

Chlorides precipitate as white silver chloride when chloride-containing water is titrated with silver nitrate solution. An indicator that provides chromate ions is potassium chromate. Silver ion concentration rises to a point at which a reddish-brown precipitate of silver chromate forms, signifying the endpoint when the concentration of chloride ions approaches extinction.

Chlorides are typically not harmful to humans, although sodium in table salt has been related to renal and heart disease. At 250 mg/L, sodium chloride may give off a salty taste, while calcium or magnesium chloride is typically not tasted until values of 1,000 mg/L are achieved. There are a number of ways that chlorides can enter surface water, including:
- chloride-containing rocks;
- runoff from agricultural operations;
- industrial wastewater;
- wastes from oil wells;
- wastewater effluent from sewage treatment facilities;
- salting of roads during snowfall.

6.1.9 Sulfates

Sulfates (SO_4^-) are found in almost all natural water sources. Polyatomic anion is used in a wide range of sectors, including pharmaceuticals, cleaning supplies, and cosmetics. While some sulfate minerals are naturally occurring in some soils and rocks, sulfate is mostly a contaminant that enters our water supply from waste and industrial discharge. Mines, smelters, paper mills, textile mills, and tanneries produce sulfate, which ends up in streams and groundwater. The main sulfates found in our water are sodium, potassium, and magnesium sulfates, which are all very soluble.

Sulfate concentrations in saltwater are around 2,700 mg/L, while freshwater lakes have concentrations ranging from 3 to 30 mg/L. The **W**orld **H**ealth **O**rganization (**WHO**) recommends a maximum sulfate content of 500 mg/L in the guidelines for drinking water quality. Sulfate levels greater than 250 mg/L may cause the water to taste bitter or medicinal. Sulfate levels beyond a certain threshold can corrode plumbing, particularly copper piping. Plumbing materials that are more resistant to corrosion, such as plastic pipes, are widely used in places with high sulfate levels. Sulfate is removed from drinking water using four different treatment systems:

- Reverse osmosis, which forces water through a membrane with microscopic pores. The membrane prevents various pollutants, including sulfates, from passing through while enabling water to pass through. Depending on the type of treatment unit, reverse osmosis eliminates between 93% and 99% of the sulfates in drinking water.
- Distillation is a method of producing steam by boiling water. As the steam rises, pollutants such as sulfates are left behind. Distillation units, when used properly, can remove nearly all sulfates.
- Anion exchange is the most commonly used method for eliminating substantial amounts of sulfates from commercial, livestock, and public water supplies. Individual residential water treatment is not often used. It is a method of replacing negatively charged ions such as sulfates with salts such as sodium chloride or potassium chloride.
- Adsorptive media filtration uses a charged media bed to pull ions with opposing charges such as sulfates out of the water and attach them to the media.

Titration with a solution of barium chloride ($BaCl_2$) or barium nitrate ($Ba(NO_3)_2$) can be used to assess the presence of sulfate ions in water. As shown in Figure 6.11, both will be sources of barium ions that react with sulfate ions in water solutions. If there are sulfate ions in the solution, a white precipitate of barium sulfate ($BaSO_4$) forms. False findings can be obtained if the sample contains carbonates, which similarly form a white precipitate. To overcome this, the sample should be titrated with barium chloride after being treated with nitric acid or hydrochloric acid.

Indirect EDTA titration can also be used to determine sulfates. The determination of sulfate concentration in water using indirect EDTA titration is a useful experiment that is simple to carry out. Excess barium chloride is added to a water sample to precipitate sulfate ions as $BaSO_4$. After that, the unprecipitated barium ions are titrated using EDTA.

The turbidimetric approach can also be used to determine sulfate ions. Under regulated conditions, sulfate ions in a water sample are transformed to a suspension of $BaSO_4$ for turbidimetric measurement.

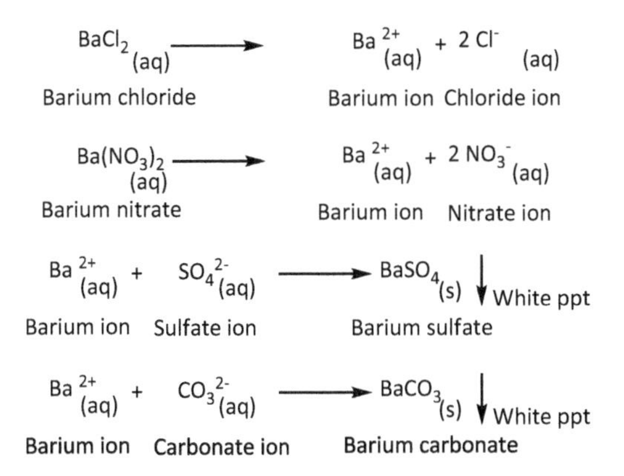

Figure 6.11: Reaction of barium ions with sulfate and carbonate ions during the determination of sulfate ions in a water sample.

To generate the suspension, the sulfate ions contained in the sample are precipitated by adding a highly acidified solution of barium chloride. A spectrophotometer is then used to calculate the suspension's percent transmittance. This is then converted mathematically to turbidity.

The turbidity of the sample solution is compared to a calibration curve constructed from standard sulfate ion solutions.

6.1.10 Dissolved Oxygen

Dissolved **o**xygen (**DO**) is used as a measure of a water body's health, with higher DO concentrations associated with good productivity and low pollution.

DO in water is measured using chemical methods such as titrimetry, electroanalytical methods (using galvanic and polarographic probes), optical DO, and colorimetric methods. Modern techniques, on the other hand, mostly employ electrochemical or optical sensor technologies.

Winkler's method is one of the most popular techniques for determining the amount of DO in water. The test involves adding manganese sulfate and alkali-iodide-azide reagent to a water sample, which results in the formation of a white precipitate of $Mn(OH)_2$. The oxygen in the water sample then oxidizes this precipitate, turning it into a brown precipitate that contains manganese. The solution is then acidified with strong sulfuric acid or hydrochloric acid, and the brown precipitate converts the iodide ion to iodine. The amount of dissolved oxygen is exactly proportionate when iodine is titrated with a thiosulfate solution.

6.1.11 Total Dissolved Solids

The elements in drinking water known as **total dissolved solids (TDS)** are both beneficial and harmful. Examples of both organic and inorganic compounds that are dissolved in water include minerals, salts, metals, cations, or anions. TDS concentrations are expressed as **parts per million (ppm)** and milligrams per liter (mg/L). TDS concentrations usually range from 50 to 1,000 ppm. The ideal TDS concentrations in the water are listed in Table 6.3.

Table 6.3: The preferable levels of TDS in water.

TDS levels in mg/L (ppm)	Rating
<300	Excellent
300–600	Good
600–900	Fair
900–1,200	Poor
>1,200	Unacceptable

However, the WHO recommends a TDS level of 300 ppm. TDS is typically reduced by distillation to less than 10 ppm and by reverse osmosis to less than 25 ppm. TDS levels in bottled mineral water typically range from 400 to 650 ppm.

Conductivity and gravimetric analysis are the two primary methods for estimating TDS. The gravimetric approach (Figure 6.12) requires letting the solvent evaporate, dry, and then weighing the leftover residues. Even if it takes some time, this experiment is usually regarded as the best. Using an electrical conductivity or TDS meter is another easy technique to determine the TDS in water and the quality of the water.

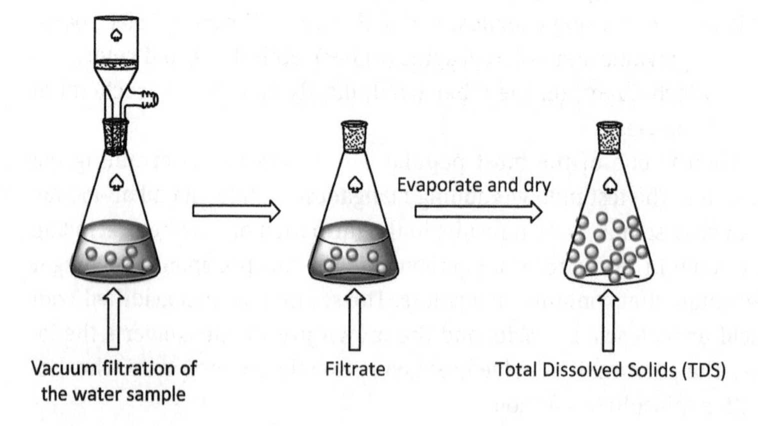

Figure 6.12: Gravimetric approach to determine the total dissolved solids.

6.1.12 Total Suspended Solids

The total number of solids dissolved in water is one way to judge the purity of water in lakes, rivers, and streams. High dissolved solids levels may indicate poor water quality. The same holds true for wastewater or drinking water that has been treated in a wastewater treatment facility.

Total **s**uspended **s**olids (**TSS**) is a term used to describe the dry weight of undissolved suspended particles in a sample of water that may be filtered through or measured by sintered glass crucible (glass fiber filter). **T**otal **s**uspended **m**atter (**TSM**) and **s**uspended **p**articulate **m**atter (**SPM**) are other terminologies used to describe TSS. The exact essential measurement is described by all three words. Previously, TSS was also referred to as **n**on-**f**ilterable **r**esidue (**NFR**).

A preweighed glass fiber filter (or regular filter) that collects the particles is used to filter a water sample to quantify TSS, as shown in Figure 6.13. The filter that was removed from the Buchner funnel after filtration is now dried in an oven for 1 h at 103–105 °C to eliminate any leftover water before being weighed once again. The TSS (the dried residues on the glass fiber filter) concentration in mg/L is given by the weight differential times 1 million over the sample volume.

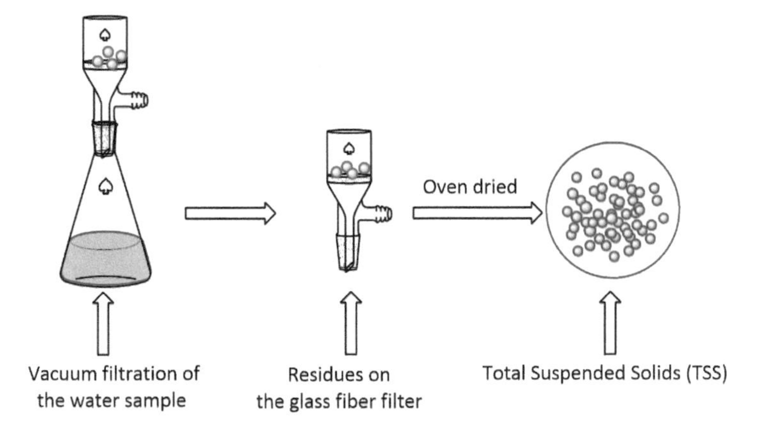

Figure 6.13: Gravimetric approach to determine the total suspended solids.

6.1.13 Coliform Bacteria Test

Municipal water treatment operators should monitor coliform bacteria levels in the water, with zero total coliform being the health requirement. However, private well monitoring is the duty of the homeowner for well water safety. Testing well water for bacteria is the only accurate way to know if water is safe because most of the time no one can identify if water contains disease-causing organisms (pathogens) based on its appearance, taste, or smell. Health officials highly advise annual coliform bacteria testing for private water wells since contamination can occur without any change in the water's flavor or odor. There are five methods or techniques for detecting the presence of coliform bacteria in water samples. They are as follows:

- Membrane filter (MF) technique
- Multiple tube fermentation (MTF) technique
- Presence–absence (P-A) method
- Defined substrate technology (DST) method (the Colilert test)
- Heterotrophic plate count (HPC) or standard plate count (SPC) method

6.1.13.1 Membrane Filter Technique

The **m**embrane **f**ilter (**MF**) techniques, shown in Figure 6.14, can be used to assess whether a specific coliform group is present in wastewater and groundwater. To determine the quality of the water, the number of colonies is directly counted using a colony counter. A bacterial colony equates to a single bacterium in a 100 mL sample of water.

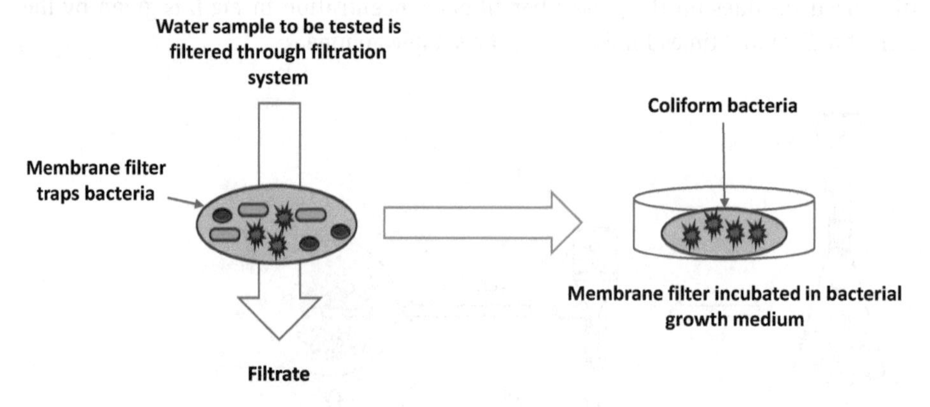

Figure 6.14: Membrane filter technique for coliform bacteria detection.

6.1.13.2 Multiple Tube Fermentation Technique

A **m**ultiple-**t**ube **f**ermentation (**MTF**) technique involves the analysis of water samples for the presence of coliform bacteria. A set of water-filled culture tubes containing lauryl tryptose broth or lactose broth is incubated at 35 °C for 48 h. Figure 6.15 shows that if coliform bacteria are present in the water sample, they will break down lauryl tryptose or lactose to produce CO_2 gas. The presence of gas indicates that the test is positive.

6.1.13.3 Presence–Absence Method

The **p**resence-**a**bsence (**P-A**) test with 4-**m**ethyl**u**mbelliferyl-β-**D**-**g**lucuronide (**MUG**) is commonly used for routine drinking water testing. Coliform will ferment lactose to produce acid and gas if it is present. In addition to lactose, P-A broth contains a pH indica-

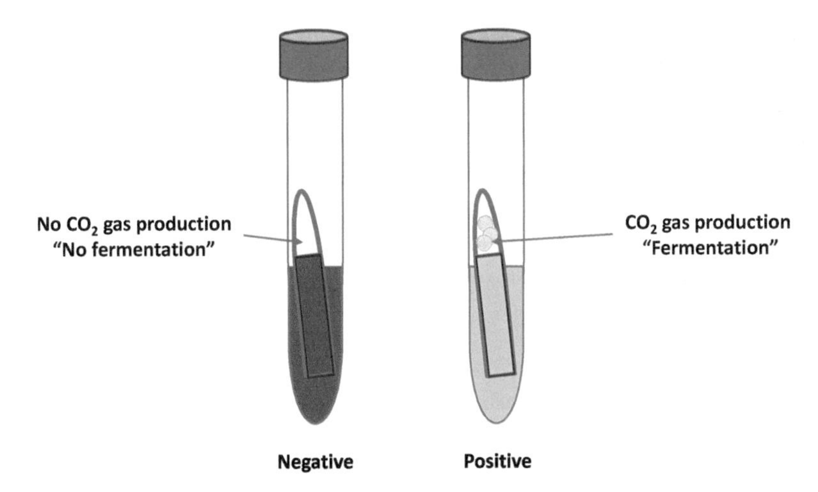

Figure 6.15: Multiple tube fermentation technique for coliform bacteria detection.

tor (bromocresol purple), which changes from purple to yellow when lactose ferments and acid is created as shown in Figure 6.16. MUG is also added to this P-A broth and is hydrolyzed by the *E. coli*-glucuronidase enzyme to generate 4-methylumbeliferone, which fluoresces under long UV light (336 nm). Detecting fluorescence emission is another sensitive method for confirming the presence of *E. coli* in water samples.

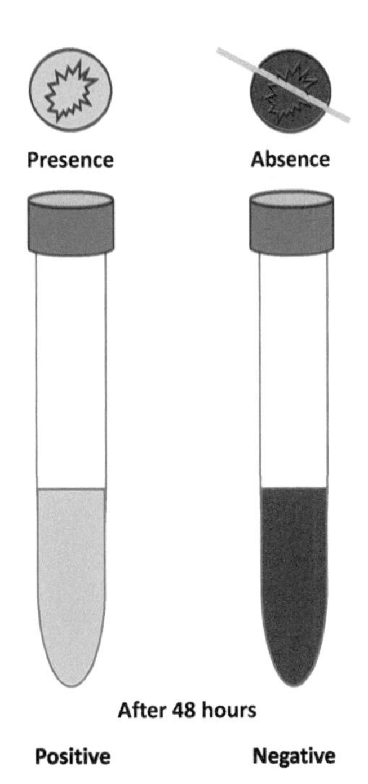

Figure 6.16: P-A test with MUG for coliform bacteria detection.

6.1.13.4 Defined Substrate Technology Method

The **d**efined **s**ubstrate **t**echnology (**DST**), also known as the Colilert test, is a reagent system used to identify and count certain target microorganisms in a bacterial mixture. The device directly counts total coliforms and *Escherichia coli* from a water sample. The reagent contains ***o*-nitrophenyl**-β-D-**g**alactopyranoside (**ONPG**; Figure 6.17A) which is digested by total coliforms to create a yellow chromogen and 4-**m**ethyl**um**beilliferyl-β-D-**g**lucuronide (**MUG**; Figure 6.17B) which is hydrolyzed and fluoresces when E. coli organisms develop.

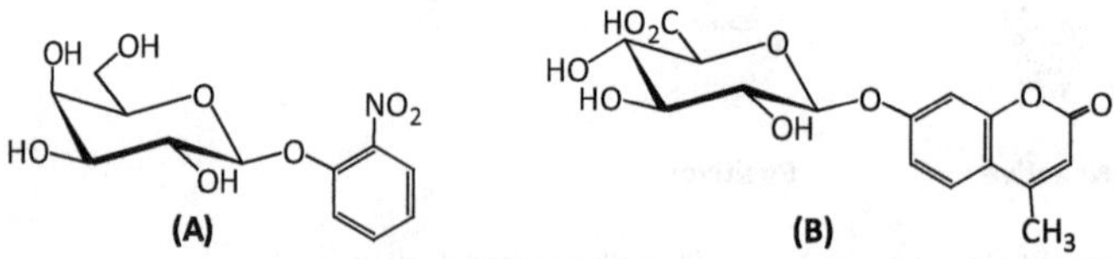

Figure 6.17: Chemical structures of ONPG and MUG.

6.1.13.5 Heterotrophic Plate Count (Standard Plate Count) Method

The **h**eterotrophic **p**late **c**ount (**HPC**), shown in Figure 6.18, or **s**tandard **p**late **c**ount (**SPC**) method is commonly used to determine the population of heterotrophic microorganisms in drinking water and other media. Heterotrophs are organisms that require an external source of organic carbon to grow, such as bacteria, yeasts, and molds. This method calculates the total number of bacteria in the water that can grow on nutrient agar as a general medium. A colony count of less than 500 per mL of the sample indicates that the water is properly disinfected, and vice versa.

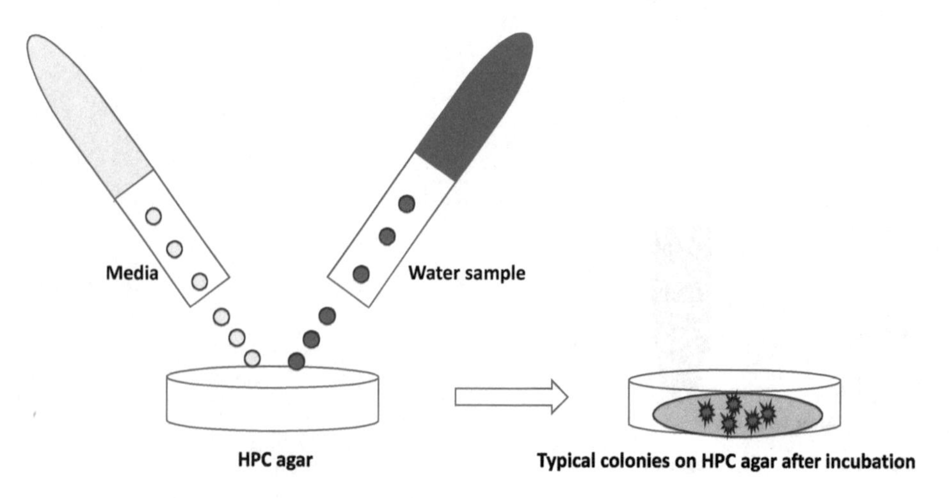

Figure 6.18: Heterotrophic plate count.

The test is performed by pouring a low-nutrient medium into a Petri plate with the sample and incubating it for 48 h. The number of heterotrophs present is then determined by counting all colonies. It should also be emphasized that the results of an HPC test are not a true estimate of overall heterotrophic concentrations, but rather are indicators of the presence of culturable organisms.

HPC limits of 100–500 colony-forming units per mL are recommended by drinking water quality guidelines around the world.

6.2 Questions

6.2.1 Give examples of tools that are used for water analysis and mention their use.

6.2.2 What does bluish-green or green mean in terms of possible pollutants?

6.2.3 What does NTU stand for?

6.2.4 What is the main goal of using the jar test?

6.2.5 Five samples are tested for alkalinity. Using the following table, determine the types of alkalinity in each sample.

- The first sample has no P alkalinity, and total alkalinity is 240 mg/L.
- The second sample has a total alkalinity of 120 mg/L and a P alkalinity of 10 mg/L.
- The third sample has 60 mg/L P alkalinity and 120 mg/L total alkalinity.
- The fourth one has 60 mg/L P alkalinity, and the total is 80 mg/L.
- The fifth sample has all-P alkalinity and total alkalinity of 95 mg/L.

Titration result	Hydroxide (OH)	Carbonates (CO_3^{2-})	Bicarbonate (HCO_3^-)
P = 0	0	0	T
P < ½ T	0	2P	T−2P
P = ½ T	0	T	0
P > ½ T	2P−T	2(T−P)	0
P = T	T	0	0

6.2.6 Orthotolidine test is used for determination of:

A. Dissolved oxygen
B. Residual chlorine
C. Biochemical oxygen demand
D. Dose of coagulant

6.2.7 Explain the difference between temporary and permanent hardness.

6.2.8 What are the two chlorine concentration levels available for the DPD comparator disks?

6.2.9 Name the reagent that is used in the chloroscope test.

6.2.10 What is the difference between TDS and TSS?

Chapter 7
Water Quality

Water is the second most important necessity for life, after air. As a result, water quality has received extensive attention in the scientific literature. Water quality is most frequently defined as "the physical, chemical, and biological properties of water." Water quality is a measurement of a body of water's condition in relation to the needs of one or more biotic species and/or any human need or purpose. Water can be contaminated by agricultural, industrial, and residential activities that produce a wide range of pollutants such as heavy metals, pesticides, fertilizers, poisonous chemicals, and oils.

The four types of water quality are potable water, palatable water, contaminated (polluted) water, and infected water. Figure 7.1 depicts the most generally used scientific definitions of different types of water quality.

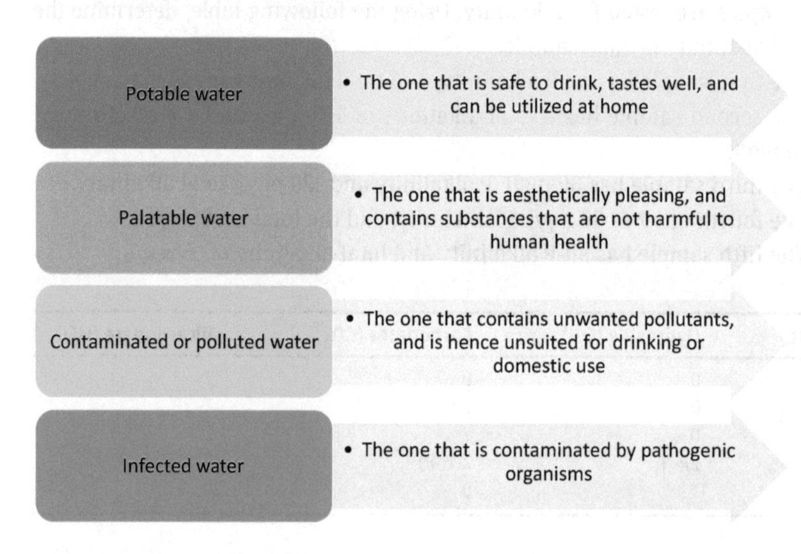

Figure 7.1: Types of water quality.

Certain toxins in our water can cause health problems such as gastrointestinal disease, reproductive problems, and neurological impairments. Infants, young children, pregnant women, the elderly, and persons with compromised immune systems may be more vulnerable to sickness. To maintain an acceptable water quality, water should be colorless and transparent, free of any aroma or smell, contain only the quantity of dissolved salts, essential for human health to give water a modest taste, and contain dissolved gases such as oxygen, which offers freshness to water. There are additionally six main water quality indicators. These are shown in Figure 7.2.

https://doi.org/10.1515/9783111332468-007

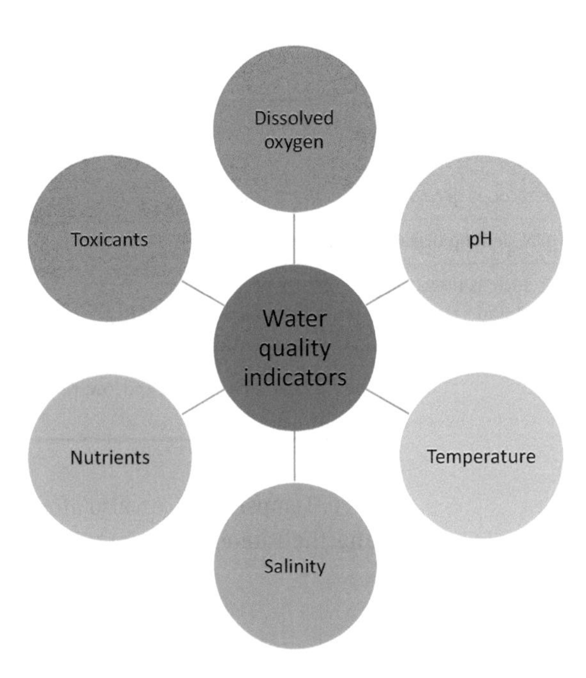

Figure 7.2: Six main water quality indicators.

7.1 Parameters of Water Quality

Water quality refers to the state of the water, including its chemical, physical, and biological qualities, as well as its suitability for a certain purpose such as drinking or cooling or swimming. These three categories of water quality metrics are discussed in greater depth in the following paragraphs.

7.1.1 Physical Parameters of Water Quality

7.1.1.1 Turbidity

The capacity of light to flow through water is measured by turbidity. It is caused by particle matters suspended in water. Turbidity in drinking water is prohibited because it gives water an unpleasant appearance. Turbidity effects are summarized in Table 7.1.

7.1.1.2 Temperature

It is one of the most essential criteria for water quality. Temperature influences palatability, viscosity, solubility, odor, and chemical reactions. As a result, temperature influences the processes of sedimentation, chlorination, and **b**iochemical **o**xygen **d**emand (**BOD**). It also has an impact on the biosorption of heavy metals in water.

Table 7.1: Turbidity effects and consequences.

Money (operational cost)	Raise the cost of water treatment for a variety of applications
Particulates	Act as a haven for hazardous germs, shielding them from the disinfection process
Suspended debris	Clog or injure fish gills, limiting disease resistance and slowing growth rates
Suspended particles	Absorb heavy metals such as mercury, chromium, lead, and cadmium, as well as many toxic organic pollutants such as **poly****c**hlorinated **b**iphenyl**s** (**PCBs**), **p**olycyclic **a**romatic **h**ydrocarbon**s** (**PAHs**), and many pesticides
Temperature	Raise water temperature, lowering the amount of accessible food and lowering the **d**issolved **o**xygen (**DO**) concentration

It is always best to keep water temperatures below 20 °C. Temperature can also affect the amount of oxygen that can dissolve in water and the rate at which algae and other aquatic plants can photosynthesize.

7.1.1.3 Color

Organic trash such as fallen leaves, twigs, and pulled weeds decay and colorize water, as do poisonous waste such as batteries, pesticide containers, medication bottles, and motor oil, and inorganic garbage such as plastic soda bottles, yogurt cups, and plastic bags. Water color should be checked to maintain good water quality by evaluating samples. This can be done by using the platinum-cobalt color scale.

The platinum-cobalt scale, commonly known as the Apha–Hazen scale, is a color scale that evaluates "yellowness" in liquids, using dilutions of a 500 ppm (parts per million) platinum cobalt solution. Its color index is a method for measuring product quality and contaminants as well as evaluating pollution levels in water or wastewater. The Platinum–Cobalt color scale runs from 0 to 500 **p**latinum **c**olor **u**nits (**PCUs**), with "0" referring to water as white or "distilled." A 500 on the scale indicates that the water has a pronounced yellow color due to pollution or breakdown products. Color values that are evaluated in comparison to the platinum–cobalt standards are denoted as PCU. The World Health Organization recommends that the color of drinking water not exceed 15 true color units (equivalent to platinum color units).

7.1.1.4 Taste and Odor

Organic materials, inorganic compounds, or dissolved gases can generate taste and odor in water. These resources could be natural, domestic, or agricultural. The quantitative value of odor or taste is assessed by measuring a volume of sample **A** and diluting it with a volume of sample **B** of odor-free distilled water until the odor of the resulting combination is hardly detectable at a total mixture volume of 200 mL. The odor or taste unit is represented in terms of a threshold number, which is as follows:

$$\text{TON or TTN} = (A + B)/A$$

(where **TON** is the **t**hreshold **o**dor **n**umber and **TTN** is the **t**hreshold **t**aste **n**umber).

7.1.1.5 Solids

Solids exist in water in two forms: solution and suspension. These two forms of solids can be distinguished by passing a water sample through a glass fiber filter. The suspended solids are held on the top of the filter, while the dissolved solids pass through with the water. If a part of the filtered water sample is placed in a tiny dish and then evaporated, the solids remain as a residue. Total dissolved solids, or TDS, is the common name for this substance:

$$\text{Total solid (TS)} = \text{total dissolved solid (TDS)} + \text{total suspended solid (TSS)}$$

Fixed solids are defined as the remnant of TSS and TDS after heating to dryness for a specified time and temperature. Volatile solids are solids that are lost when heated to 550 °C. These measurements are useful to wastewater treatment plant operators because they generally estimate the amount of organic matter present in total solids of wastewater, activated sludge, and industrial wastes. They are determined as follows:

$$\text{Total solids(mg/L)} = [(\text{TSA} - \text{TSB})] \times 1,000/\text{sample(mL)}$$

where TSA is the weight of dried residue + dish in milligrams and TSB is the weight of dish in milligrams

$$\text{Total dissolved solids(mg/L)} = [(\text{TDSA} - \text{TDSB})] \times 1,000/\text{sample (mL)}$$

where TDSA is the weight of dried residue + dish in milligrams and TDSB is the weight of dish in milligrams

$$\text{Total suspended solids (mg/L)} = [(\text{TSSA} - \text{TSSB})] \times 1,000/\text{sample (mL)}$$

where TSSA = weight of dish and filter paper + dried residue and TSSB is the weight of dish and filter paper in milligram

$$\text{Volatile suspended solids (mg/L)} = [(\text{VSSA} - \text{VSSB})] \times 1,000/\text{sample(mL)}$$

where VSSA = weight of residue + dish and filter before ignition (mg) and VSSB = weight of residue + dish and filter after ignition (mg).

7.1.1.6 Electrical Conductivity (EC)

Water's **e**lectrical **c**onductivity (**EC**) is a measure of a solution's capacity to carry or conduct electrical current. Because electrical current is carried by ions in solution, conductivity rises as ion concentration rises. As a result, it is one of the primary factors used to establish water appropriateness for irrigation and firefighting. It is measured in the following units:

$$\text{U.S. units} = \text{micromhos/cm}$$

$$\text{S.I. units} = \text{milliSiemens/m (mS/m) or dS/m (deciSiemens/m)}$$

where $(\text{mS/m}) = 10\,\mu\text{mho/cm}\ (1{,}000\,\mu\text{S/cm} = 1\,\text{dS/m})$.

Pure water is not a good conductor of electricity. Typical conductivity of water is as follows:
- Ultrapure water: 5.5×10^{-6} S/m
- Drinking water: 0.005–0.05 S/m
- Seawater: 5 S/m

7.1.2 Chemical Parameters of Water Quality

7.1.2.1 Biochemical Oxygen Demand

Biochemical oxygen demand (BOD) is the quantity of dissolved oxygen that aerobic biological organisms in a body of water need to break down organic material present in a certain water sample at a specific temperature during a specific time period. The faster that oxygen is lost in the stream, the higher the value. Higher-level aquatic life thus has less access to oxygen.

7.1.2.2 Phosphorus

Phosphorus, one of the five essential components of living things, is required for organic life. Although elemental phosphorus is uncommon in nature, it does exist in a number of forms, most notably as phosphates. The soluble form of phosphate is orthophosphate (PO_4^{3-}), a naturally occurring ion in water. Excess phosphorus in surface water should be avoided because aquatic plants and algae can multiply quickly and produce a range of water quality problems, including low dissolved oxygen levels, which can kill fish and other aquatic life.

7.1.2.3 pH

In almost every water quality application, pH testing is critical. pH is regulated in wastewater treatment as part of discharge permits, and many treatment procedures are pH-dependent. High or low pH readings in environmental sampling and monitoring can indicate contamination. Pollution can alter the pH of water, causing harm to aquatic animals and plants. The following summarizes the effects of pH:
- Most aquatic creatures and plants have adapted to survive in water with a specific pH and may be adversely affected by even minor changes.
- Even moderately acidic (low pH) water can reduce the quantity of hatched fish eggs, irritate the gills of fish and aquatic insects, and damage membranes.

- Water with an extremely low or high pH is lethal. Most fish will die if the pH falls below 4 or rises above 10, and very few creatures will survive if the pH falls below 3 or rises above 11.
- Low pH endangers amphibians because their skin is particularly sensitive to pollutants. Some scientists believe that the present decline in amphibian populations around the world is caused by low pH levels caused by acid rain.
- Heavy metals like cadmium, lead, and chromium dissolve more quickly in very acidic (lower pH) water and become significantly more hazardous.
- A change in pH can alter the forms of several compounds in water. Ammonia, for example, is relatively innocuous to fish in neutral or acidic water. However, when the water becomes more alkaline (the pH rises), ammonia becomes more toxic to these same creatures.

7.1.2.4 Total Organic Carbon

Total **o**rganic **c**arbon (**TOC**) is a metric used to determine how much carbon is present in all organic compounds in both pure water and aqueous systems. Businesses and labs can assess a solution's compatibility, with their procedures, by using this valuable analytical technique known as TOC.

Organic pollution can impair ion exchange capacity and encourage undesirable biological growth, rendering treated water unfit for human consumption. Some harmful by-products could be generated through water treatment.

7.1.3 Biological Parameters of Water

7.1.3.1 Bacteria

Bacteria proliferate so quickly that counting the number of bacteria in a sample of water is nearly impossible. In most circumstances, bacteria reproduce slowly in cooler water. Many dangerous waterborne infections, such as cholera, tularemia, and typhoid, can be caused by high levels of bacteria in water.

7.1.3.2 Blue-Green Algae

Blue-**g**reen **a**lgae (**BGA**), also known as cyanobacteria, can be found in a variety of colors, including blues, greens, reds, and black. BGA can reduce nitrogen and carbon in water, but it can also deplete dissolved oxygen when present in large quantities. Monitoring BGA is critical because algal blooms pose a major danger to water quality, ecological stability, surface drinking water supplies, and human health due to toxin production and the massive biomass produced.

7.1.3.3 Nutrients

Nitrogen is a nutrient that occurs naturally in both fresh and salt water. Plant development in aquatic ecosystems is depedent on it. When substantial volumes of nitrogen are introduced into river ecosystems, problems develop. As a result, excessive algal growth may occur, reducing available oxygen in streams used by fish and other aquatic life.

7.1.3.4 Viruses

Due to their small sizes, viruses are able to pass through the majority of filters. Certain waterborne viruses can cause hepatitis and similar health problems. Despite the difficulty in treating viruses, most water treatment facilities should be able to eliminate viruses during the disinfection process.

7.2 Question

7.2.1 What is the TON, if 170 mL of odor-free distilled water is required to produce a 200 mL mixture from a 30 mL odor?

7.2.2 What is the distinction between potable and palatable water?

7.2.3 Describe three turbidity effects and their consequences.

7.2.4 How and in what manner is the color of drinking water evaluated?

7.2.5 What do TTN and TON mean?

7.2.6 What is ultrapure water's typical conductivity?

7.2.7 Define BOD.

7.2.8 What is the water-soluble form of phosphate?

7.2.9 What happens if the pH of the water falls below 4?

7.2.10 List the six most important water quality indicators.

Chapter 8
Water Treatment

8.1 Drinking Water

Treatment for drinking water production entails removing pollutants and/or inactivating potentially hazardous bacteria from raw water to produce water that is pure enough for human use without any risk of ill health effects in the short or long term.

Boiling is one of the oldest and most effective methods for killing disease-causing organisms such as viruses, bacteria, and parasites. The high temperature and length of time spent boiling are critical for effectively killing the organisms in the water. Even if the water is still hazy or murky, boiling will effectively cure it. Another method of water purification is reverse osmosis (RO). RO filters are the most successful at eliminating pollutants from water, including potentially hazardous microorganisms.

On a large scale, raw water treatment involves screening, coagulation and flocculation, aeration, sedimentation, filtration, and disinfection. The full operation is depicted in Figure 8.1.

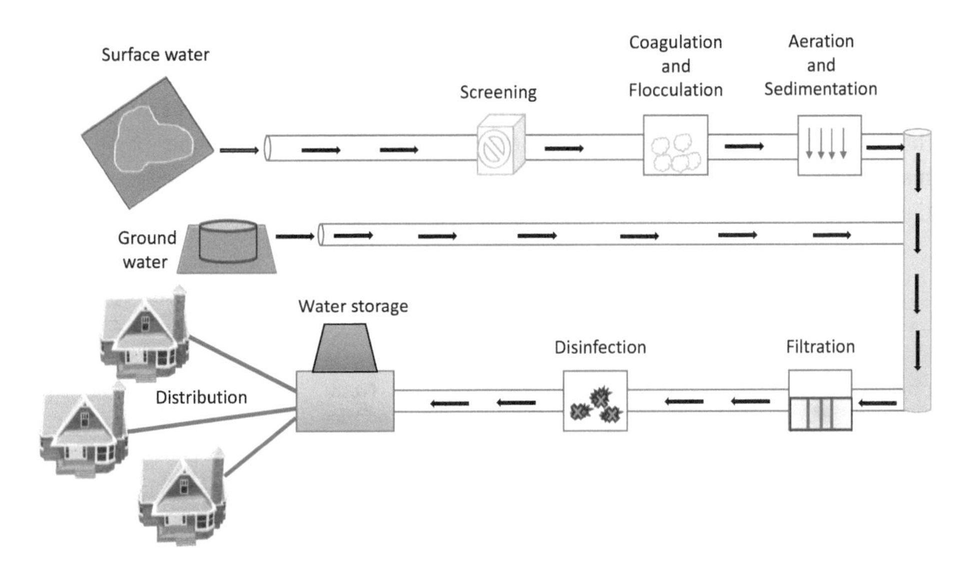

Figure 8.1: Large-scale raw water treatment.

Small-scale water treatment systems commonly comprise one or more of the five major methods described below:

- Disinfection by chlorination and **u**ltra**v**iolet (UV) light
- Activated carbon filtration
- RO

https://doi.org/10.1515/9783111332468-008

- Distillation
- Ion exchange

8.2 Seawater

Desalination is an artificial process that converts saline water (usually seawater) to fresh water by eliminating dissolved mineral salts. When applied to seawater, this procedure is currently one of the most often employed to obtain fresh water for human consumption or agricultural uses. Desalination can be accomplished in three ways:

8.2.1 Distillation (Desalination)

Distillation is a common method to desalinate seawater or ocean water to obtain fresh water by condensation of vapors produced by heating as simply illustrated in Figure 8.2. Thermal desalination systems, in general, employ heat to evaporate water, leaving behind dissolved components. After that, the condensed water vapor is collected as product water. Distillation is the most fundamental of these thermal processes, and its energy efficiency has significantly improved. **M**ultistage **f**lash (**MSF**) distillation is the most widely used thermal desalination method nowadays. MSF reuses waste heat by cycling it through a series of low-pressure chambers. Another low-pressure chamber-based thermal process is **m**ultiple-**e**ffect **e**vaporation (**MEE**), also known as **m**ultiple-**e**ffect **d**istillation (**MED**).

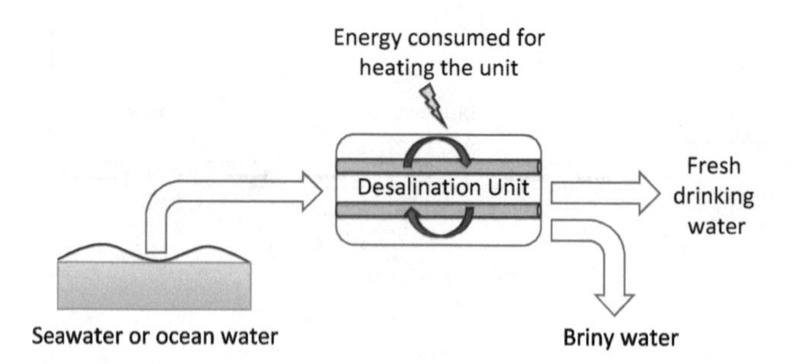

Figure 8.2: Seawater thermal desalination process.

8.2.2 Reverse Osmosis

Reverse **o**smosis (**RO**) is a method of obtaining fresh water that involves filtering seawater under pressure via a semipermeable membrane, as illustrated in Figure 8.3. In

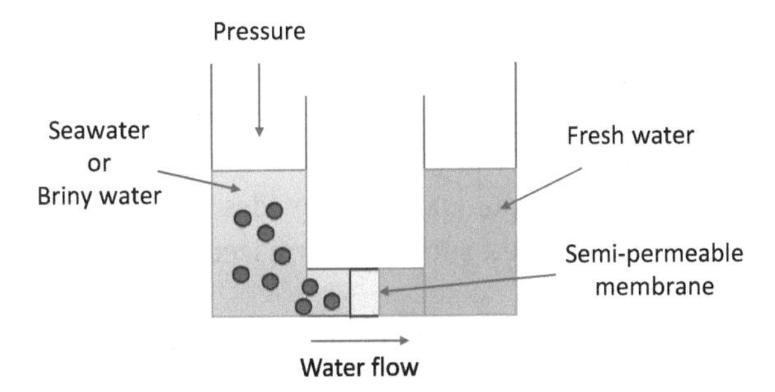

Figure 8.3: Reverse osmosis.

this most popular membrane desalination process, water molecules are forced through very small pores (holes) under high pressure, while salts and other larger molecules are preserved.

8.2.3 Electrodialysis

Electrodialysis (ED) extracts ions from water by using current. Figure 8.4 depicts a technique of desalinating water that does not require the use of filters. Instead, an electrically generated shockwave is employed to replace filters, resulting in cheaper and easier-to-manage drinking water.

In this approach, briny water flows through a frit, a porous substance comprised of glass particles that are sandwiched between membranes or electrodes on either side. When an electric current is passed through the system, the salinized water separates into zones with either enriched or depleted salt concentrations. When the current is raised, the fresh and salty sections of the streams are separated by a simple physical barrier at the flow's center, causing a shockwave to form between the two zones and splitting the streams.

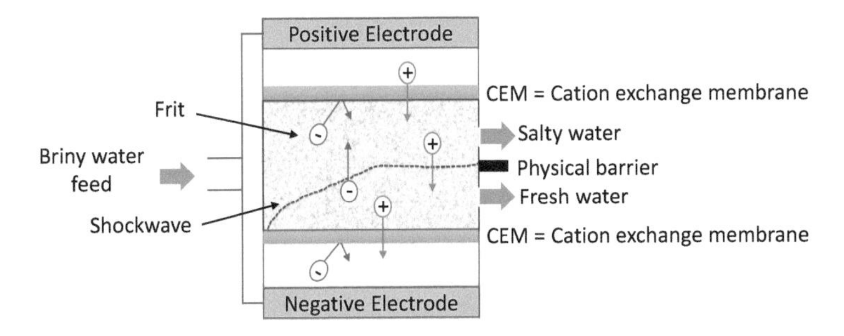

Figure 8.4: A sample shock electrodialysis diagram.

8.2.4 Advantages of Seawater Desalination

Desalination can offer water to places that are possibly water-stressed or desert. Desalination procedures can enable access to previously unusable saline waters in numerous situations. Due to the high quality of the output water, it provides safe drinking water. It can also offer water for other businesses, such as pharmaceuticals, which require very pure water sources.

Desalination has the potential to considerably increase climate change adaptation by diversifying water suppliers and increasing tolerance to water quality degradation. Water supply diversification can provide other or supplemental sources of water when current water resources are insufficient in quantity or quality. Desalination technologies are also resistant to water quality degradation since they may produce extremely pure water from highly contaminated source waters.

8.2.5 Disadvantages of Seawater Desalination

The primary drawbacks of present desalination techniques include costs, energy requirements, and environmental repercussions. The disposal of the concentrated waste stream, as well as the effects of intakes and outfalls on nearby ecosystems, is among the environmental consequences. In addition, the high energy requirements of the current desalination systems will raise greenhouse gas emissions and may obstruct climate-change mitigation efforts.

8.3 Municipality and Industrial Water Waste

Municipality and industrial water waste treatment are designed to improve wastewater quality by removing pollutants and toxicants, protecting water quality of natural water resources, and preventing harmful diseases by eradicating pathogenic organisms found in wastewater.

8.3.1 Sources

- Domestic sanitary wastewater is wastewater produced by human and household wastes, such as kitchen sinks, bathtubs, toilets, showers, and laundry.
- Industrial wastewater contains hazardous chemicals and other pollutants from industries, manufacturers, and mills.
- Commercial wastewater comes from schools, hospitals, workplaces, hotels, restaurants, and airports.
- Stormwater is wastewater that drains from roofs, parks, gardens, streets, and gutters.

8.3.2 Methods of Treatment

Wastewater treatment technology or advanced wastewater treatment methods can be classified into three types:

- Physical treatment: Physical removal of pollutants and contaminants.
- Chemical treatment: Chemical reactions are employed to eliminate impurities or hazardous pollutants.
- Biological treatment: Pollutants are expelled from the environment through biological activity.

8.3.3 Treatment Processes

Industrial and municipal wastewater treatment processes use four unique steps of treatment to eliminate hazardous contaminants in order to generate clean effluent that can be properly discharged to water bodies.

8.3.3.1 Preliminary Treatment

Preliminary treatment removes large debris, coarse particles, and heavy inorganic material from the wastewater flow. It consists of physical operations such as:

- Screening: This process removes heavy materials in wastewater in order to reduce downstream equipment damage and blockage.
- Floatation is a method of separating floatable and suspended solid particles from wastewater.
- Grit removal: Grit chambers are used to slow down the flow of water and allow sediments to settle out of it so that they can be removed manually or mechanically.

8.3.3.2 Primary Treatment

Primary treatment involves the use of a sedimentation tank or primary clarifier that removes most of the suspended solids that will float or settle. Sedimentation often uses chemicals like flocculants and coagulants. The resulting primary sludge that settles to the bottom of the clarifier is collected for a further treatment called sludge treatment.

8.3.3.3 Secondary Treatment

Secondary treatment is a biological approach that treats effluent further to remove leftover organics and suspended particles. Secondary biological treatment procedures are classified as aerobic (with oxygen) or anaerobic (without oxygen). This treatment involves an activated sludge process in which the primary effluent is routed via an aeration tank where the air is mixed with sludge, followed by a secondary sedimentation tank where microorganisms and solid wastes clump and settle. In general, this treatment removes pollutants as well as a significant portion of dangerous compounds.

8.3.3.4 Tertiary or Advanced Treatment

Tertiary or advanced treatment is the ultimate step in wastewater processing, and its major goal is to remove certain wastewater constituents that cannot be removed in previous stages, thus increasing the effluent quality to a larger degree. It uses filtration to remove more suspended particles than primary and secondary screening and sedimentation can.

Nutrients, heavy metals, certain dangerous chemicals, and other contaminants and impurities are removed through this method. It requires disinfection, which can be performed with both physical disinfectants like **u**ltra **v**iolet (**UV**) radiation and chemical disinfectants like chlorine. During this process, a substantial percentage of dangerous organisms are killed or regulated.

8.4 Questions

8.4.1 What is the purpose of drinking water treatment?
8.4.2 What are the most ancient and successful strategies for eliminating disease-causing organisms?
8.4.3 Describe the steps involved in large-scale raw water treatment.
8.4.4 What are the most often used small-scale water treatment methods?
8.4.5 What is the most common method of thermal desalination?
8.4.6 What does MSF stand for?
8.4.7 What is RO?
8.4.8 Is filtering essential during the ED process?
8.4.9 List the three advanced wastewater treatment methods.
8.4.10 What physical procedures are employed in the preliminary treatment of wastewater?

Chapter 9
Water and Wastewater Treatment Techniques

9.1 Coagulation

Coagulation is the use of coagulants and coagulant aids to precipitate colloidal turbidity particles. In order to precipitate out the negatively charged colloidal turbidity particles, positively charged inorganic coagulants are used.

When inorganic coagulants such as aluminum sulfate (alum), ferric chloride, and ferric sulfate are introduced to water, coagulation occurs, which is crucial to the treatment process because it lessens electrostatic repulsion between particles. The negatively charged nonsettable materials that are scattered throughout the water are balanced by the positive charge of the inorganic coagulant, forcing the water's particles to coalesce. The coagulation process is shown in Figure 9.1, in which negatively charged colloidal turbidity particles are drawn in by Al^{3+} or Fe^{3+} cations (+ve ions) aggregation, resulting in the formation of micro-flocs, a microscopic clumps.

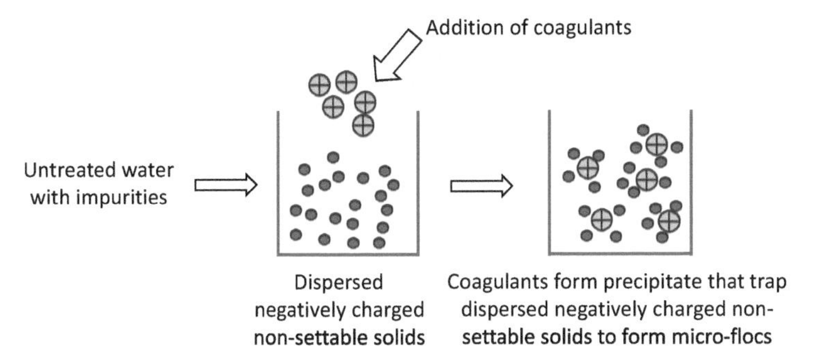

Figure 9.1: Coagulation process.

The type of coagulant employed and a variety of other variables, including those listed in Table 9.1, affect the size of flocs formed during the coagulation process. The majority of microplastics, other particles, and dissolved detritus may be removed from water via coagulation.

9.1.1 Coagulants in Water Treatment

Aluminum and iron-based coagulants are used in water treatment to remove particles such as microplastics, TOC, phosphorus, BOD and COD, color, metals, hydrogen sulfide, conditioning sludge, dewatering sludge, bulking sludge, and lake restoration. Table 9.2 lists some of the most commonly used coagulants.

https://doi.org/10.1515/9783111332468-009

Table 9.1: Factors affecting coagulation.

pH	The effectiveness of the coagulant generally depends upon pH. The optimum pH range for precipitation with aluminum salts is between 5 and 7, while iron salts can be used in the range 5–11.
Alkalinity	Anions like OH^- are required to generate insoluble compounds, which could then be precipitated out.
Temperature	Reaction and coagulation occur more quickly and effectively at higher temperatures.
Time	Detention time and proper mixing are crucial.
Velocity	Lower velocity will allow the floc particles to settle in the flocculation basin, whereas higher velocity will shear or break them.

Table 9.2: Aluminum and iron-based coagulants.

Aluminum-based coagulants	Iron-based coagulants
Aluminum sulfate	Ferric chloride
Polyaluminum sulfate	Ferrous chloride
Aluminum chloride	Ferric sulfate
Polyaluminum chloride	Ferrous sulfate
Aluminum chlorohydrate	Ferric chlorosulfate
Sodium aluminate	

9.1.2 Coagulant Aids

A coagulant aid facilitates coagulation by improving coagulation conditions such as pH and alkalinity. Lime and sodium carbonate are the two pH-adjusting coagulant aids. Clay, sodium silicate, and activated silica are non-pH-altering coagulant aids. Polymers are coagulating aids that function as secondary coagulants.

They are added to water during coagulation to improve coagulation conditions such as pH and alkalinity as well as make a more stable and stronger floc, solve the issue of slow floc generation in cold water, reduce the amount of coagulant that is required, and reduce the amount of sludge that is produced.

9.2 Flocculation

Flocculation is the separation of a solution, commonly the removal of sediment from a fluid. As shown in Figure 9.2, flocculation occurs when silt clumps into bigger aggregated particles flakes that are easier to recognize and remove. This process can happen naturally or be generated by using flocculants or physical approaches. In the water treatment process, the flocculation phenomenon is used to remove particles from water by producing larger clusters or flocs. This is a common method of treating stormwater, wastewater, and purifying drinking water.

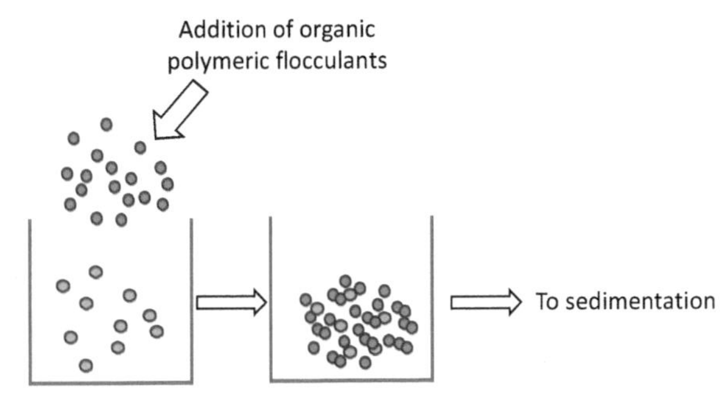

Figure 9.2: Flocculation process.

9.2.1 Flocculants

Flocculants are substances that cause small particles in a solution to clump together, resulting in the development of a floc that floats to the surface or sinks to the bottom. This can then be removed more easily from the liquid. Flocculants come in a wide range of charges, charge densities, molecular weights, and morphologies. Inorganic flocculants, organic synthetic polymer flocculants, natural polymer flocculants, and composite flocculants are the four types of flocculants. Table 9.3 shows some examples of these categories.

Table 9.3: Types of flocculants.

Inorganic flocculants	**Organic synthetic polymer flocculants**
Aluminum sulfate Sodium aluminate Ferrous sulfate	Polyacrylamide Sodium polyacrylate, polyoxyethylene Polyvinylamine Polyvinyl sulfonate
Natural polymer flocculants	**Composite flocculants**
Starch phosphate Starch Xanthate Chitosan Chitin-based flocculants	Polyacrylamide ferric chloride Polysilicate aluminum sulfate Polyaluminum ferric chloride Polyaluminum chloride-chitosan

Organic polymeric flocculants are the most widely utilized today due to their ability to produce flocculation with minimal dosage. However, because of their lack of biodegradability and the concomitant dispersion of potentially toxic monomers into water systems, researchers are shifting their focus to more environmentally friendly biopolymers. The disadvantage is that they have a shorter shelf life and a higher dosage requirement than organic polymeric flocculants. To tackle these drawbacks, integrated solutions are being developed in which synthetic polymers are grafted onto natural polymers to create customized flocculants for water treatment that provide the best of both worlds. Flocculants can be used alone or in conjunction with coagulants. This depends on the chemical composition of the water to be purified. The type of contaminants removed from the water as well as the method of separation utilized by the water treatment facility define the optimal combination of organic and inorganic flocculants and coagulants.

9.3 Sedimentation

The separation of flocs from flocculated water by gravity is known as sedimentation. This physical method of wastewater purification is typically carried out in sedimentation tanks or basins. Inlet and output valves generally control water flow in these basins or tanks, and the bottom floor slopes to a hopper for sludge collection.

The efficacy of this method is determined by particle size and weight. While heavier particles fall, suspended solids, with specific gravity similar to water, remain suspended. Sedimentation benefits municipalities by using fewer chemicals for later water treatment and simplifying any subsequent process. It is less expensive than other technologies, and the water quality that passes through the process varies less.

A simplified sedimentation diagram is shown in Figure 9.3. Sediments or sludge collect in the bottom, as shown, and can be readily removed. Coagulants are also often used in water to help with settling, prior to sedimentation.

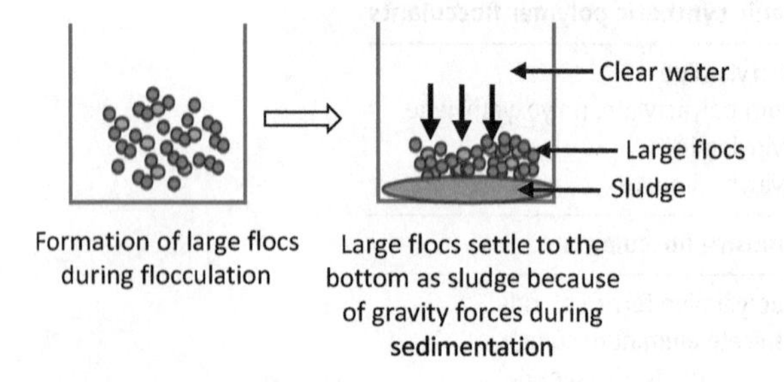

Figure 9.3: Sedimentation process.

There are several types of sedimentation basins (*also known as sedimentation clarifiers*). The most common ones are conventional basins, high-rate basins, and solid contact basins. These are covered in greater depth in the following paragraphs.

9.3.1 Conventional Basins

Simplified diagrams of rectangular and circular basins are depicted in Figure 9.4. Rectangular sedimentation tanks have length-to-width ratio of at least 4 to 1. In order to ensure proper sedimentation, sludge, light floc (above the sludge), and clear water are kept well separated on top. From the intake to the outlet, the rectangular basin has a horizontal flow.

The circular basin receives radial flow from a central intake, with inlets and outflows at opposite ends. Circular tanks with diameters of up to 60 m are in use but are usually limited to 30 m to prevent wind effects. In the circular sedimentation tanks, the influent is routed through the tank's central pipe, resulting in a radial flow. Mechanical sludge scrappers are installed to collect the sludge, which is then conveyed through the sludge pipe at the bottom.

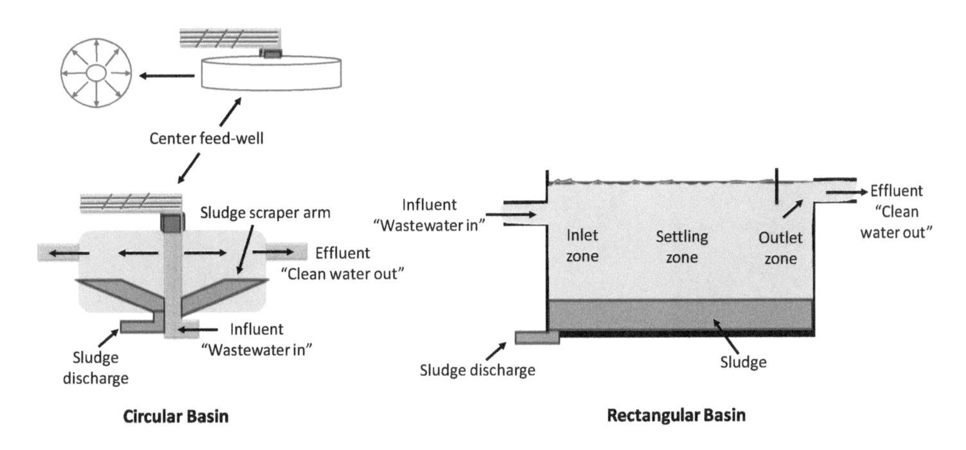

Figure 9.4: Rectangular and circular basins.

9.3.2 High-Rate Basins

High-rate basins are intended for improved drinking water and wastewater treatment with a high load and minimal detention time. They are compact units. The average detention duration is 1–2 h, compared to 4–6 h in typical basins. These basins are divided into three types: tube settler basins, plate settler basins, and solid contact basins.

The tube settler basins, as shown in Figure 9.5, have tubes installed in them to increase the settling surface and offer sufficient baffling for greater sedimentation in less space. Tube settlers combine a number of tubular channels that are sloped at a 60° angle and close to one another to give a larger effective settling area. This results in a significantly lower particle settling depth than typical basins, resulting in shorter settling times.

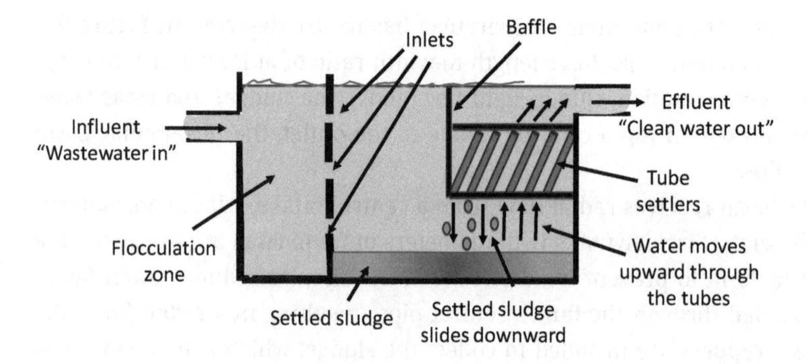

Figure 9.5: Tube settler basin.

Plate settler basins, depicted in Figure 9.6, serve the same purpose as tubes but use plates instead of tubes. Plate settlers improve the effective settling area by using a series of inclined plates, commonly made of stainless steel, and placed two to three inches apart on a 55° to 60° slope.

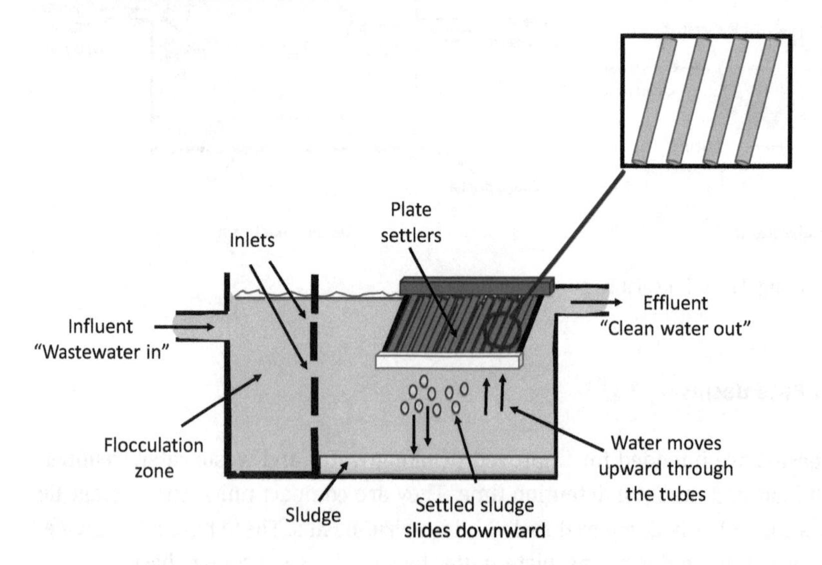

Figure 9.6: Plate settler basin.

A solid contact basin, depicted in Figure 9.7, is a small device that combines zones of rapid mixing, coagulation, and sedimentation. In a single basin, they use the recycled solids contact principle for flocculation, sedimentation, and clarifying. The influent water is combined with previously settled sediments within the draft tube. The moderate mixing of the process encourages the agglomeration of floc particles and chemical precipitates. The aggregated materials settle out quickly in the clarifying area. Rotating sludge scrapers transport the settled sediments to the basin's center for disposal.

Solid contact clarifiers are used in big industrial operations such as refineries and petrochemical facilities to produce process water that can be used as a general process or as cooling water throughout the plant. Solid contact clarifiers can also be used in municipal drinking water systems in place of traditional flocculation and sedimentation basins to reduce the environmental impact.

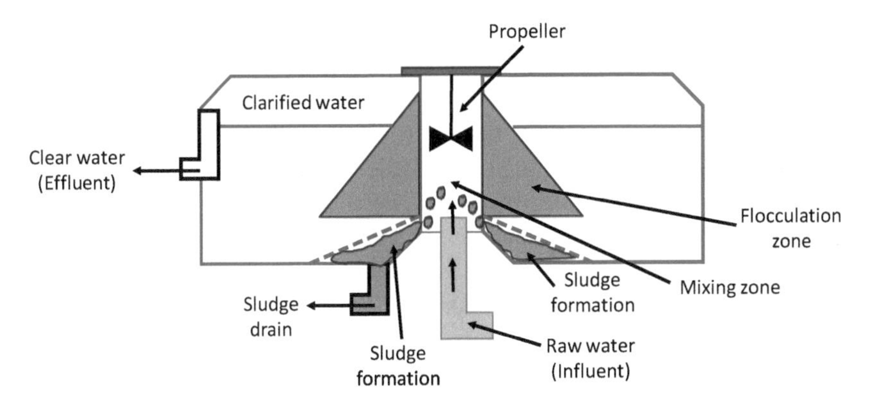

Figure 9.7: Solid contact basin.

9.3.3 Sedimentation Influencing Factors

Different elements, including the following, influence sedimentation:

- Sufficient detention time is essential for full sedimentation. The turbidity increases and the settling decreases as the detention period decreases, and vice versa.
- Proper velocity is crucial. At greater velocities, the settled floc is resuspended, which may cause it to rise to the surface and result in excessive effluent turbidity. The solids in the basin would not be distributed properly at a relatively low velocity.
- The rate of sedimentation decreases with increasing turbulence, and vice versa.
- The rate of sedimentation increases with temperature, and vice versa. Water settles more quickly and is lighter at higher temperatures.

- The basins' precise dimensions, particularly their depth, which is often 15 feet or more, are crucial.
- The horizontal flow basins' inlets and outlets need to be positioned correctly to allow for proper water mixing and prevent short circuits.
- Chemical feed points should be carefully selected to ensure that each chemical reacts properly with its target ingredient, that is, alum should always be used before lime.
- Appropriate sludge removal is necessary for managing the decomposition, scouring, and effective volume of the basin. Denser sludge settles faster and needs to be removed more frequently than lighter, frothier, or flakier types.

9.4 Softening

Water softening is a process of removing calcium, magnesium, and iron ions from water. Substances containing additional positively charged ions are more difficult to dissolve in hard water because of the presence of these ions. The role of this method of water softening is to eliminate undesirable minerals from the water.

The use of ion exchange (I-EX) and/or the addition of chemicals that cause insoluble precipitates are two methods for softening water. On a small scale, ammonia, borax, calcium hydroxide (slaked lime), and trisodium phosphate, in combination with sodium carbonate (soda ash), are used for softening. To remove the precipitates, sedimentation and filtration must come after the lime-soda method of water softening. By adding just enough lime to precipitate the calcium as the carbonate and the magnesium as hydroxide, followed by sodium carbonate to eliminate the remaining calcium salts, water can be chemically softened on a large scale.

By passing the water through columns of a natural or synthetic resin that exchange sodium ions for calcium and magnesium ions, I-EX is often done. In the following paragraphs, we will go into more detail about those softening techniques as well as others such as chelating agents, distillation, and reverse osmosis (RO).

Water softening makes it possible to:

- Reduce the amount of hard water deposits or stains in your baths and showers
- Have simple and quick household cleaning
- Lower energy expenses due to greater water heater efficiency
- Put on whiter, brighter, and softer apparel
- Extend the life of water-using equipment

9.4.1 Softening Methods

9.4.1.1 Chemical Precipitation Methods

Lime softening: It is a water treatment method that softens water by eliminating calcium and magnesium ions using calcium hydroxide (limewater, $Ca(OH)_2$). Hydrated lime is added to the water during this process to raise the pH and precipitate the ions that produce hardness. Lime addition removes only magnesium hardness and calcium carbonate ($CaCO_3$) hardness.

That means it will act on $CaCO_3$ hardness and also magnesium carbonate and magnesium non-carbonate hardness, such as magnesium sulfate, as shown in Figure 9.8.

$$Ca(HCO_3)_2 + Ca(OH)_2 \longrightarrow 2\,CaCO_3\downarrow + 2H_2O$$

$$Mg(HCO_3)_2 + 2\,Ca(OH)_2 \longrightarrow Mg(OH)_2\downarrow + 2\,CaCO_3\downarrow + 2\,H_2O$$

$$MgSO_4 + Ca(OH)_2 \longrightarrow Mg(OH)_2\downarrow + CaSO_4\downarrow$$

Figure 9.8: Lime softening reactions.

Soda (soda ash, Na_2CO_3) softening: It is a treatment process that turns calcium and magnesium chlorides and sulfates into their corresponding carbonates, which precipitate as shown in Figure 9.9.

$$CaSO_4 + Na_2CO_3 \longrightarrow CaCO_3\downarrow + Na_2SO_4$$

$$MgSO_4 + Na_2CO_3 \longrightarrow MgCO_3\downarrow + Na_2SO_4$$

$$MgCl_2 + Na_2CO_3 \longrightarrow CaCO_3\downarrow + 2\,NaCl$$

Figure 9.9: Soda softening reactions.

Due to the poor solubility of $CaCO_3$ and $Mg(OH)_2$, lime soda softening cannot produce water that is completely devoid of hardness. Thus, the lowest attainable calcium hardness is around 30 mg/L as $CaCO_3$, while the lowest achievable magnesium hardness is around 10 mg/L as $CaCO_3$.

9.4.1.2 Nonprecipitation Methods

There are various nonprecipitation methods for softening water. Here are five examples:

I-EX softening: The I-EX resin technique is one of the most used softening procedures. It softens water by replacing calcium and magnesium ions with sodium ions (and maybe anions for chloride ions) in sodium chloride (Figure 9.10). It can also demineralize water by substituting H^+ ions for cations and OH^- ions for anions.

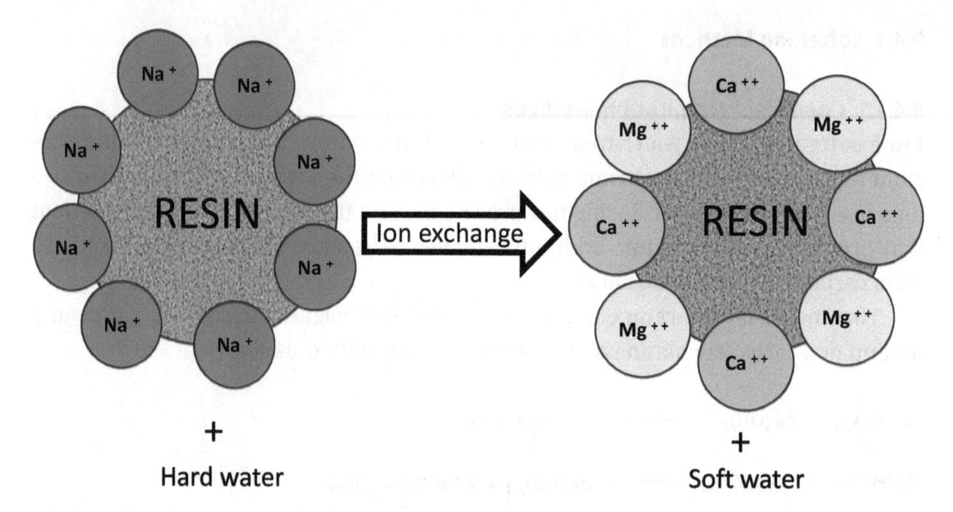

Figure 9.10: Ion exchange.

I-EX is done by flowing water through a resin column that exchanges sodium ions for calcium and magnesium ions as shown in Figure 9.11. If calcium and magnesium begin to appear in the water departing the column, cleaning the exchanger bed with water and column regeneration by slowly running a concentrated solution of common salt through are required.

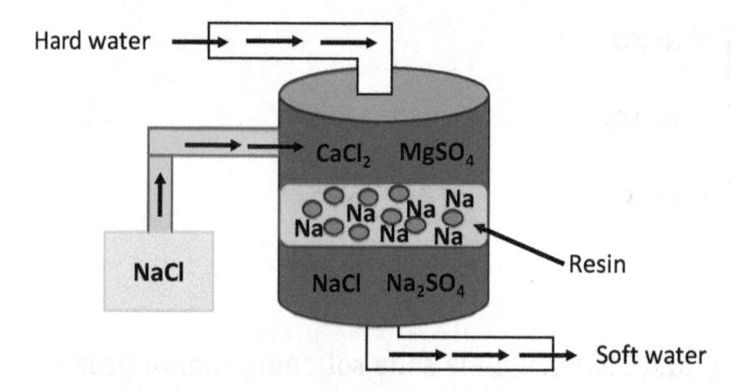

Figure 9.11: Ion exchange process.

The **dem**ineralization **(DM)** plant is a common example of the I-EX process. By this technique, deionized water is produced by combining cation and anion resins.

This is performed in two stages, with raw water first passing through a column containing a strong cation resin and then passing through a column having a strong anion resin.

DM or the removal of dissolved mineral solids by an I-EX process is based on the foundations of an I-EX reaction, in which minerals and salts dissociate into their constituent ions in the presence of water. These dissolved solids contain anions (negatively charged ions) and cations (positively charged ions), both of which are attracted to counterions (or ions with opposing charges).

A resin formed of plastic beads with an ionic functional group bonded to them is used in an I-EX column. These functional groups loosely retain ions with opposing charges by mutual electrostatic attraction. Two-bed or dual-bed ion exchangers or mixed-bed ion exchangers can be used to accomplish water DM.

A two-bed (also known as dual-bed) ion exchanger, shown in Figure 9.12, treats water with two I-EX resin beds or columns, each containing a different type of I-EX resin. Water is initially treated with a **s**trong **a**cid **c**ation **(SAC)** resin, which collects dissolved cations and releases hydrogen (H^+) ions in return. The mineral acid solution is then transferred to the resin bed for the **s**trong **b**ase **a**nion **(SBA)**. This second step removes dissolved anions while also producing water by releasing hydroxide (OH^-) ions, which combine with hydrogen (H^+) ions.

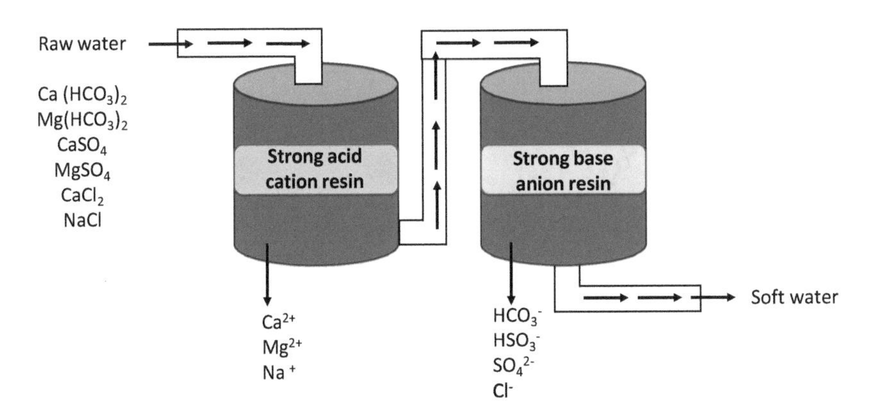

Figure 9.12: Two-bed or dual-bed ion exchanger.

A mixed-bed ion exchanger has multiple I-EX resins accommodated in a single I-EX column, as depicted in Figure 9.13. When water enters the unit, the cation and anion exchange activities occur simultaneously.

Zeolite softening: Zeolites are microporous crystalline aluminosilicates that absorb water and other cations, filling the micropores. They are a category of minerals that comprise aluminosilicates, which are produced by AlO_4 and SiO_4 tetrahedrons linked by oxygen vertices and contain interchangeable positive ions (cations) such as Na^+, K^+, Ca^{2+}, and Mg^{2+} (Figure 9.14).

Zeolites contain the chemical formula $M_{2/n}O{\cdot}Al_2O_3{\cdot}xSiO_2{\cdot}yH_2O$, where n is the valence of the charge-balancing non-framework cation M, x is 2.0 or greater, and y is the mole of water in the voids. For the water softening process, sodium-hydrated aluminosilicates ($Na_2O{\cdot}Al_2O_3{\cdot}xSiO_2{\cdot}yH_2O$) are utilized, which can reversibly exchange sodium ions with the cations in the water.

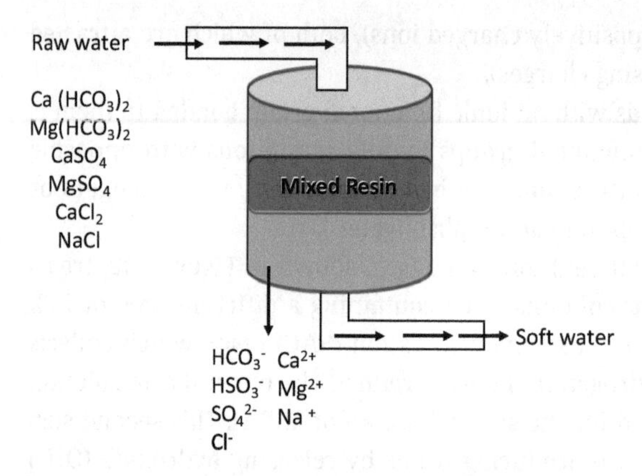

Figure 9.13: Mixed-bed ion exchanger.

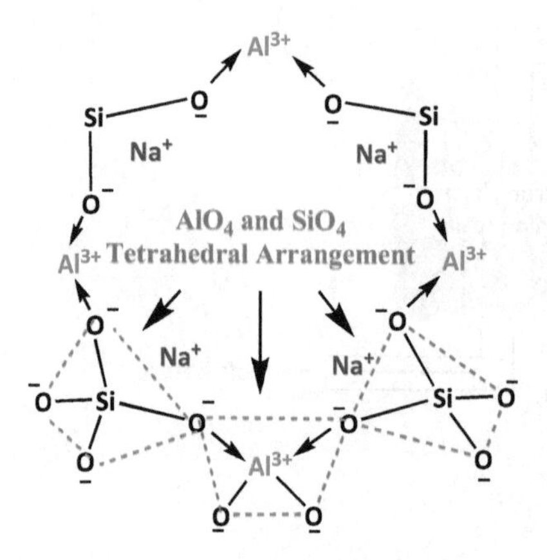

Figure 9.14: Sodium zeolite chemical structure.

When hard water is allowed to stand in sodium zeolite, Ca^{2+} and Mg^{2+} ions replace Na^+ ions, softening the water that comes out of this column and making it free of hardness-causing ions. When all of the exchangeable Na^+ in the zeolites (Z) are exchanged with hardness-causing ions in the water, such as Ca^{2+} and Mg^{2+}, it becomes depleted and must be regenerated by treating with a concentrated NaCl solution (brine), as shown in Figure 9.15.

Chelating agent softening: Chelating agents are organic compounds that may extract ions from aqueous solutions and convert them to soluble complexes. Calcium disodium ethylenediamine tetraacetic acid ($CaNa_2EDTA$, Figure 9.16A) is the most commonly used chelating agent. It is an EDTA (ethylenediamine tetraacetic acid) derivative. Another example is the biodegradable natural chelating agent fulvic acid (Figure 9.16B), which is the most powerful natural chelating agent.

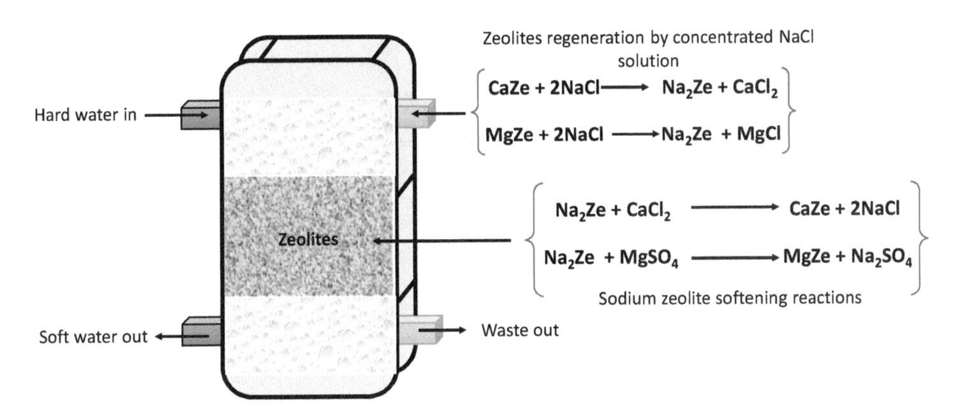

Figure 9.15: Sodium zeolite softening and regeneration.

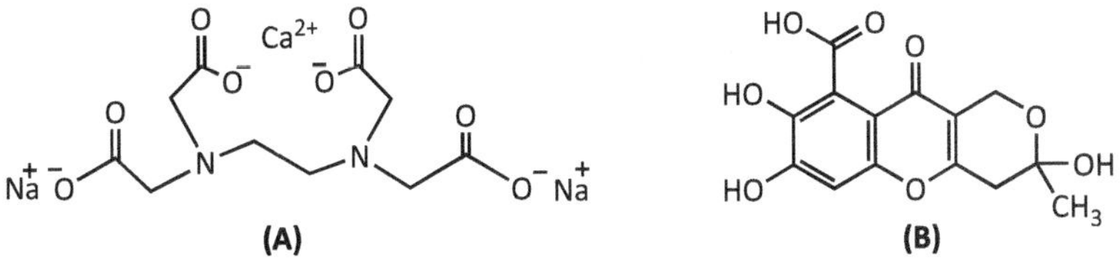

(A) (B)

Figure 9.16: Chelating agents.

Distillation softening: Distillation is the separation of the components of a liquid mixture by evaporating the liquid at its boiling point and condensing its vapors. Separation is accomplished through the difference in vapor pressures (volatilities) of the various components. It can be used to soften water by extracting large amounts of dissolved particles, such as calcium and magnesium, from it during evaporation as shown in Figure 9.17.

Membrane softening: Membrane filtration, such as **RO**, softens water while removing any remaining particles. Figure 9.18 depicts how a RO system removes contaminants and dissolved minerals from water by forcing them through a filter (membrane). RO systems can help by eliminating the source of the stench. It is more environmentally friendly than other water filtration systems because it does not require chemicals. It can also eliminate scents and unusual colors caused by contaminants in water and dissolved minerals.

9.5 Stabilization

Water stabilization is a method of decreasing and controlling water contamination. It entails lowering the amount of contamination entering bodies of water via physical,

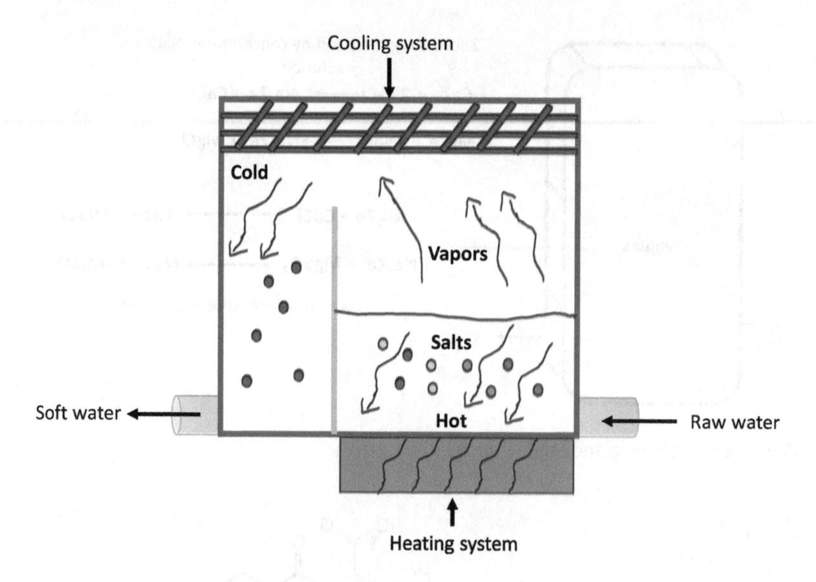

Figure 9.17: Water softening by distillation.

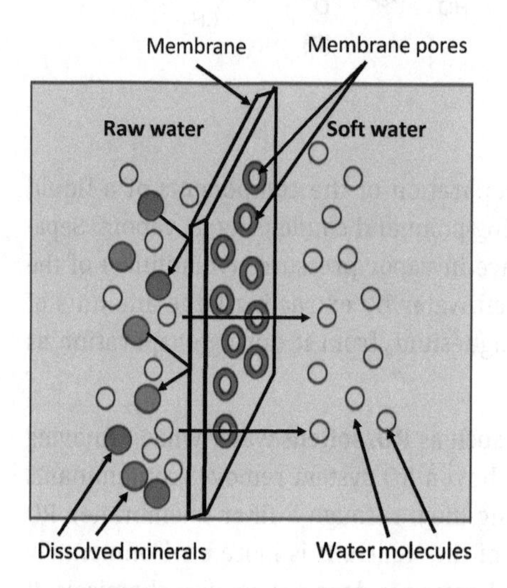

Figure 9.18: Water softening by reverse osmosis.

chemical, and biological mechanisms. These toxins can be found in agricultural runoff, sewage discharge, industrial waste, and stormwater runoff. Water stabilization works to keep water safe for human consumption and aquatic ecosystems well-being, and ensures that the treated water is not corrosive or toxic. Before delving into how various indicators and indices reflect the degree of $CaCO_3$ deposition, it is necessary to understand the terms corrosion, scaling, and depositing.

9.5.1 Scale Formation or Deposition

Scale formation on surfaces is typically divided into two distinct processes: a "deposition process" that refers to the process of heterogeneous nucleation and growth at surface asperities, and an "adhesion process" that refers to the sticking of preexisting crystals that have nucleated in the bulk.

Scale deposition is the buildup of various elements within undesirable locations. This process is depicted in Figure 9.19, which involves mineral ion clustering, nucleation, crystal development, flocculation, and depositing.

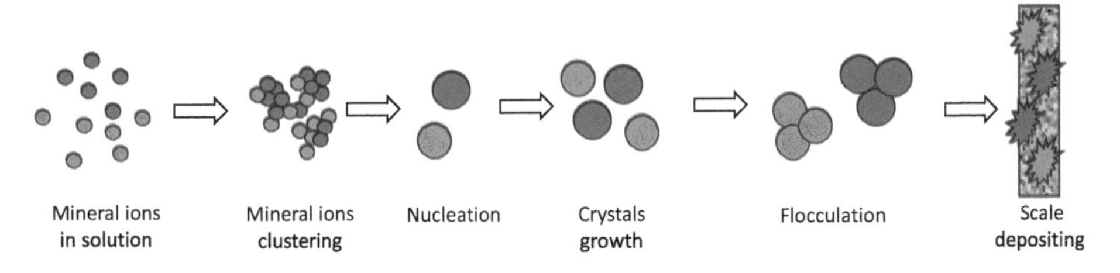

Figure 9.19: Scale deposition process.

9.5.2 Corrosion

Corrosion is an electrochemical process that predominantly transforms metals and alloys into oxides, hydroxides, and aqueous salts. During the rusting process, two reactions occur. In one reaction, the anodic reaction, metal atoms are ionized and flow into solution, but their electrons stay on the original metal surface. In the second, the cathodic step, the metal's freed electrons are taken up by chemical species such as O_2 and H_2O in the reduction processes. The simplified chemical reactions and the rusting process are depicted in Figure 9.20.

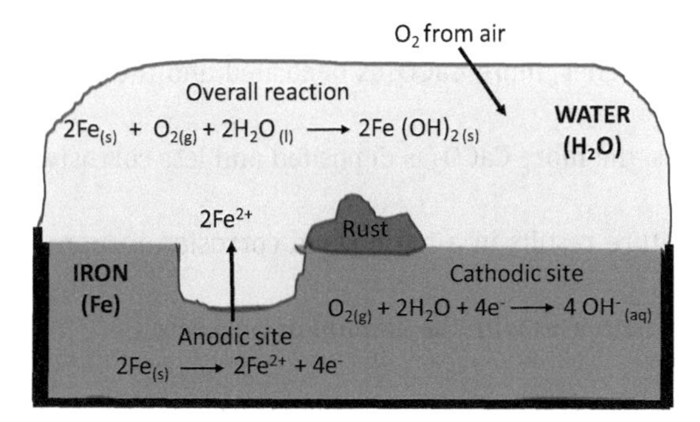

Figure 9.20: Simplified chemical reactions and the rusting process.

9.5.2.1 Corrosion Types

There are various types of corrosion, each with its own set of effects and causes. Table 9.4 summarizes some of them.

Table 9.4: Corrosion types.

Physical corrosion	Pipe surface corrosion, caused by the high velocity of particulate particles or distributed gas bubbles. It happens when two different metals in physical or electrical contact are immersed in a shared electrolyte (such as salt water) or when a metal is exposed to different electrolyte concentrations.
Galvanic or bimetallic corrosion	Happens when two different metals, such as lead and copper or stainless steel and zinc, are linked in water lines.
Stray current corrosion	Generated through direct current distribution lines, railroad systems, substations, and alternating current sources; these currents then cause corrosion when they pass through steel structures or piping systems.
Localized corrosion or pitting corrosion	Regarded as one of the most damaging types of corrosion; it begins at a single location and progresses to the production of a corrosion cell, surrounded by a regular metallic surface. It begins when an area-coated metal surface becomes exposed as a result of poor coating, corrosion, or stress. This region becomes anodic and produces a pit, while the surrounding region acts as a cathode. The pit expands as rusting progresses.
Bacterial corrosion or microbial corrosion	Microbiologically produced corrosion that is caused by organisms such as sulfur-oxidizing bacteria, iron-reducing bacteria, sulfate-reducing bacteria, and acid-producing bacteria.
Stress **c**orrosion **c**racking (**SCC**)	The metal's cracking, as a result of its corrosive surroundings and the tensions exerted on it; it frequently happens in hot environments.

9.5.2.2 Factors Affecting Corrosion

- pH: A lower pH indicates a higher level of acidity and a higher rate of corrosion.
- Alkalinity: When the alkalinity is higher, more $CaCO_3$ is deposited and the water is less corrosive.
- Hardness: The more the hardness, the more $CaCO_3$ is deposited and less corrosive is the water.
- Temperature: A greater temperature results in a more rapid corrosion response and rate.
- Dissolved gas: The corrosion rate increases with the amount of gas present.

9.5.3 Sequestration or Sequestering

Sequestration is the ability to create a complex with metal ions that allows these metal ions to remain in solution in the presence of precipitation agents (Figure 9.21). This is an extremely useful and one-of-a-kind water softening feature that provides water stabilization by mixing with hardness ions naturally found in water supplies such as calcium, magnesium, iron, and manganese.

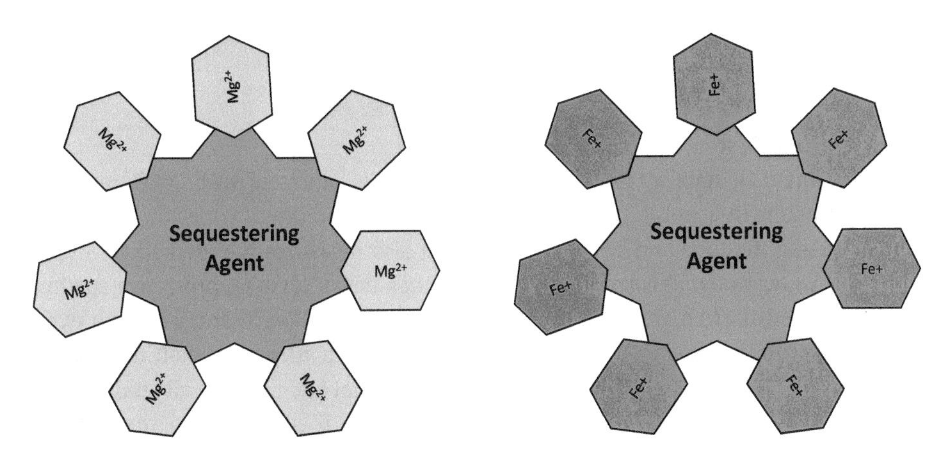

Figure 9.21: Sequestering or sequestration.

One use of sequestration is the regulation of groundwater iron and manganese without their elimination. Because chlorination of water that contains low amounts of iron results in the development of insoluble iron oxide or iron chlorides, sequencers are normally added to groundwater at the wellhead or pump intake before it comes into contact with chlorine to prevent the formation of "red water." Soluble manganese compounds react similarly to generate "black water."

Sequestering chemicals can also be used to lower hardness in water. They combine with calcium, magnesium, and other heavy metal ions in hard water to build molecules that bind the ions so tightly (sequester them) that they can no longer react.

Tetrasodium phosphate (tetrasodium pyrophosphate, Figure 9.22A), sodium hexametaphosphate (Figure 9.22B), and disodium EDTA (Figure 9.22C) are common sequestering agents.

9.5.4 $CaCO_3$ Deposition Determination

Aside from visual inspection of pipe sections and filter media, the degree of $CaCO_3$ deposition is determined by a number of indicators and indices. They are as follows:

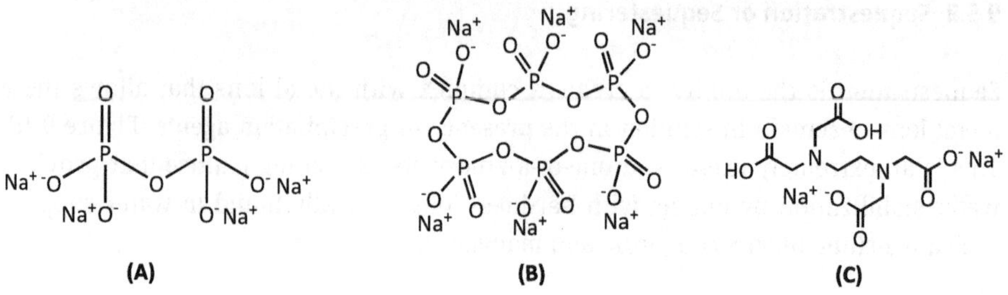

(A) (B) (C)

Figure 9.22: Sequestering agents.

9.5.4.1 Marble Test

Marble test ($CaCO_3$ stability test) determines if water is saturated with $CaCO_3$, and has two applications in general. It is useful for testing the stability of water that has been softened using the lime and soda ash method. It also measures the total alkalinity required by water to avoid corrosion by the development of a $CaCO_3$ protective coating. The marble test is often performed by soaking the water in $CaCO_3$ and leaving it overnight. The sample's alkalinity is evaluated both before and after saturation. The nature of water is determined by the difference between original and ultimate alkalinity. If the difference is greater than zero, the water is depositing; if equal to zero, the water is stable; and if less than zero, the water is corrosive.

9.5.4.2 Baylis Curve

Utilities strive to keep the water balanced in order to keep it from becoming too caustic or creating too much scale. This is performed by balancing the water's pH and alkalinity to produce stable, noncorrosive water. Some alkalinities in the form of hydroxides, carbonates, and bicarbonates should remain in water to absorb acids and prevent corrosive conditions.

Alkalinity is measured in laboratories as $CaCO_3$ and is an indicator of calcium hardness. The Baylis curve, shown in Figure 9.23, is one approach for checking if the pH and alkalinity are adequately adjusted to provide stable, noncorrosive water. Water in the top zone causes excessive scale; water in the lower zone is corrosive; and water in the equilibrium (stability) zone contains enough alkalinity to stabilize, but not enough to induce scaling.

9.5.4.3 Langelier Index

The Langelier index or Langelier saturation index **(LSI)** is an unbiased measure of water balance. This index helps to identify when water becomes corrosive or scale-forming. As shown in Figure 9.24, the ideal saturation is 0.00 LSI, and the allowed range is −0.30 to +0.30 LSI.

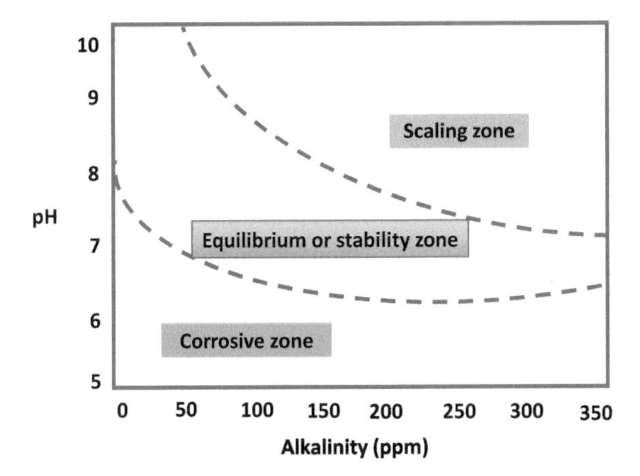

Figure 9.23: Baylis curve.

Water is aggressive if the LSI is −0.31 or lower because it is undersaturated with $CaCO_3$. The water is desperate for calcium and will go to any length to obtain it. Above +0.31, the water contains too much dissolved $CaCO_3$, causing it to precipitate $CaCO_3$. Carbonating scale, plaster dust, or other types of $CaCO_3$ could arise. Perfectly balanced water has an LSI of zero (0.00).

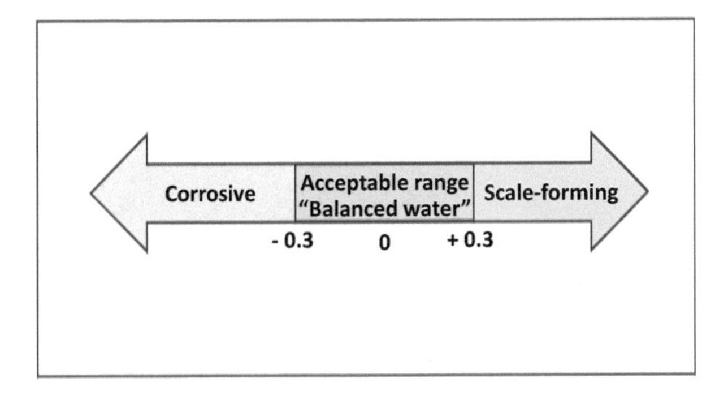

Figure 9.24: Langelier saturation Index.

9.5.4.4 Ryznar Stability Index

The **R**yznar **S**tability **I**ndex (**RSI**) is an improved version of the LSI. It employs the same argument. 2 × pHs-pH, where pHs denotes the saturation pH. In this index, 6.0 is neutral; above 6.0, there is an increasing corrosion tendency; below 6.0, there is an increasing scaling tendency. The RSI correlates higher values with more corrosive water and lower values with water that is potentially scaling. Its value range is also different from that used by the Langelier saturation index. The Ryznar index is oriented toward protecting equipment's metallic surfaces and it is designed to allow a slight coating of calcium to accumulate on metallic equipment, where it acts as a protective insulator.

9.6 Filtration

Filtration is the mechanical removal of turbidity or suspended particles from water by passing it through a porous material, such as a granular bed or a membrane.

9.6.1 Granular Media (Bed) Filtration

Granular **m**edium **f**iltration (**GMF**) is a popular physical method for removing particles from wastewater. Water travels through the granular material, while suspended solids are retained during granular filtering.

Granular media filtration enhances clarity by eliminating particles, ranging in size from coarse sediment to 10.0 μm.

The filter bed is a rectangular concrete structure containing sand or a combination of sand, garnet, and activated carbon deep media. A layer of gravel supports the medium. A drainage system for filtered water is installed beneath the gravel, as shown in Figure 9.25.

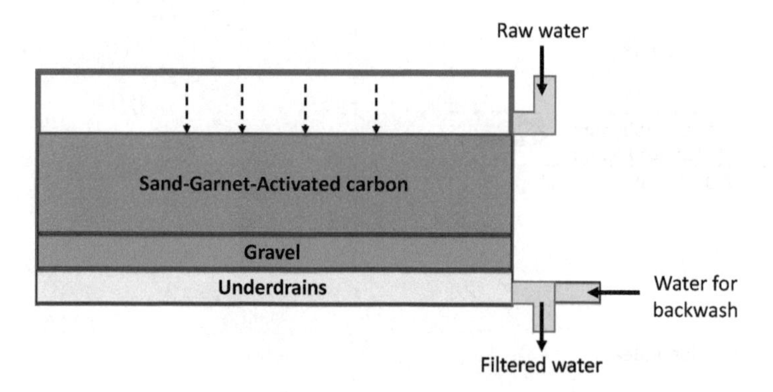

Figure 9.25: Granular media filtration.

9.6.1.1 Types of Granular Filters

- Slow sand filters
- Rapid sand filters
- High-rate sand filters
- Granular activated carbon multimedia filters
- Pressure filters

9.6.1.2 Factors Affecting the Granular Media Filtration

- The lower the turbidity in the filter influent, the longer the filter will last and function better.
- The coarser the media, the less head loss and the longer the run, and vice versa.
- The deeper the bed, the better the filtration.
- Improper washing can result in media loss, media mixing, and the development of mud balls and crackers. All of these will result in insufficient filtration.
- The higher the loading, the shorter the filter runs and the less efficient it is.
- The higher the temperature, the better the performance.
- A greater polymer dose creates fissures in the filter mat, while a smaller amount does not form an efficient micro-floc.

9.6.2 Membrane Filtration

Membrane filtration is the process of passing pretreated water under pressure through a membrane to remove particles of a certain size.

Membranes are made of cellulose acetate and synthetic materials such as polypropylene, and are either hollow tiny fibers or spiral wound structures.

9.6.2.1 Particle Removal Mechanism

- Sieving: Removes all particles larger than the pore size.
- Selective diffusion: Only certain dissolved particles travel through the membrane.
- Charge repulsion: Direct electric current is used in filtration. Anions move to the anode, while cations go to the cathode.

9.6.2.2 Types of Membranes

The four categories of membranes used in membrane filtration are shown in Figure 9.26 and their usages are briefly addressed in Table 9.5.

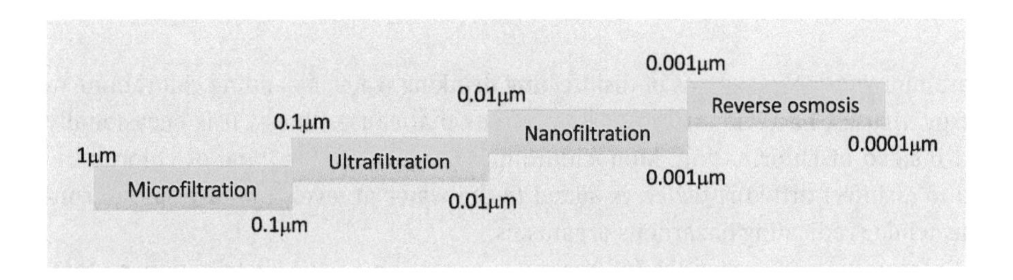

Figure 9.26: Membrane types and pore sizes.

Table 9.5: Filtration membrane types and uses.

Microfiltration	Removes all particles larger than 1 µm as well as all bacteria
Ultrafiltration	Removes all particles larger than the pore size of the filter, including viruses and other pathogens
Nanofiltration	Removes virtually all viruses, dissolved solids such as inorganic salts, and tiny organic molecules, and is frequently used to soften water and groundwater
Reverse osmosis	Removes common chemical contaminants, including sodium, chloride, copper, chromium, and lead

9.7 Disinfection

Disinfection is the eradication of water-borne microorganisms that enter the source water via sewage and runoff from the watershed. Disinfection is accomplished through the use of chemicals such as chlorine, chloramine, or chlorine dioxide. Other methods of disinfection, including ozonation, ultraviolet radiation, and photocatalytic disinfection, can be used.

9.7.1 Chlorination

Chlorination is the most used form of disinfection since it is inexpensive, effective, and useful in reducing taste and odors. It is the most commonly used disinfectant in water treatment around the world. It can be used directly as a gas or as a concentrated solution in water. It creates hypochlorous acid, a potent germicide that works by inactivating the enzymes of microorganisms. A simple water chlorination system is depicted in Figure 9.27.

9.7.2 Chloramination

Chloramination is the process of disinfecting drinking water by adding chloramine to destroy viruses, bacteria, and other organisms that cause illness. It is occasionally used instead of chlorination. Monochloramine (Figure 9.28), a type of chloramine used to disinfect drinking water, is added to the water at levels that are safe to consume, while eradicating hazardous organisms.

Chloramines are essential for residual protection in the distribution system. Water with up to 4 milligrams per liter (mg/L) of chloramine, or 4 parts per million (ppm), is safe for drinking, bathing, cooking, and other home applications. Chloramine, like other disinfectants, has pros and cons. Table 9.6 summarizes these findings.

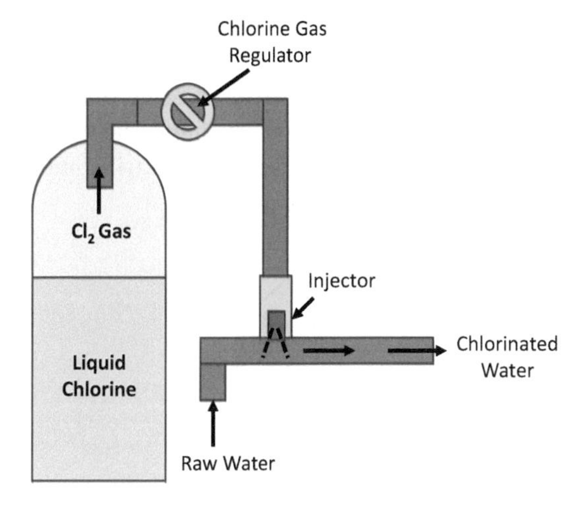

Figure 9.27: A simple water chlorination system.

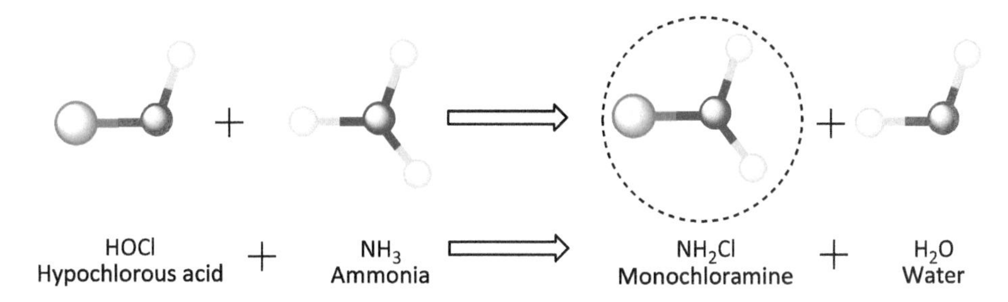

Figure 9.28: Monochloramine formation.

Table 9.6: Pros and cons of chloramine, as water disinfectant.

Pros	Cons
Chloramine is more stable, lasting longer in water, offering longer-lasting protection against dangerous germs.	To generate monochloramine on-site, chlorine gas or hypochlorite, as well as safety procedures, are required.
Because chloramine has a low oxidation potential, it does not form harmful disinfection by-products.	To prevent ammonia vaporization and the creation of nitrogen trichloride, further safety precautions are required.
In large-scale applications, chloramine is simple to utilize and is economical.	As a disinfectant against pathogens, chloramine is less effective than chlorine, necessitating longer contact durations with water to get the same results.
In water, chloramine does not have a significant chemical taste or odor.	Metal pipes are more corroded by chloramine.

9.7.3 Chlorine Dioxide

Chlorine dioxide (ClO_2, Figure 9.29) is a versatile, broad-spectrum biocide that has been used safely and effectively in the treatment of drinking water, industrial water, and wastewater. It selectively oxidizes microorganisms while emitting no trihalomethanes, bromates, or other potentially harmful disinfection by-products.

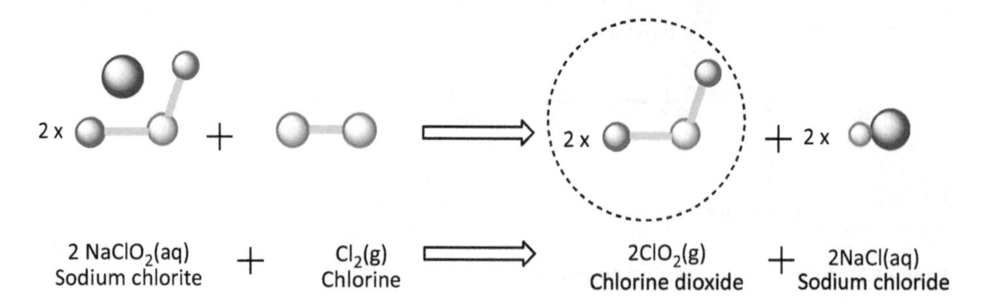

Figure 9.29: Chlorine dioxide formation.

9.7.4 Ultraviolet Light Water Treatment

UV treatment, shown in Figure 9.30, is used in the treatment of wastewater, drinking water, and aquaculture. UV light kills bacteria by altering their biological components, specifically the chemical connections in nucleic acids and proteins. UV treatment is less effective in high-turbidity water, but more successful in water with fewer dissolved solids and organic materials.

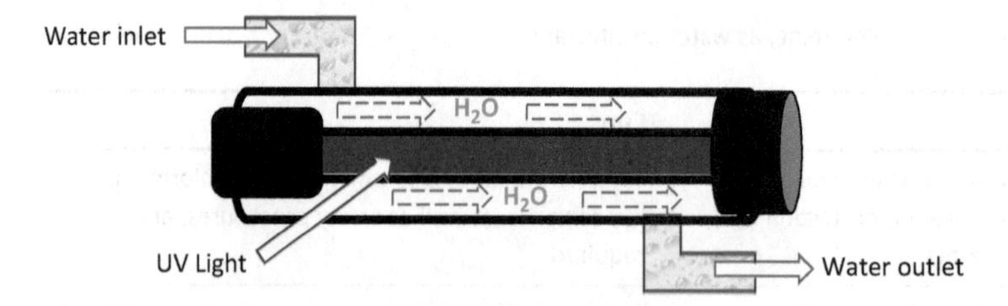

Figure 9.30: Ultraviolet light water treatment.

9.7.5 Ozonation

Ozone is produced using either ambient air or pure oxygen. The gas (oxygen) is then converted into ozone using an electric field. Adding ozone to the contact tank (Figure 9.31) and letting water dissolve it completes the disinfection process. By oxidizing the bacterial cell walls and causing them to disintegrate (lyse), ozone kills bacteria. Ozone is a nontoxic

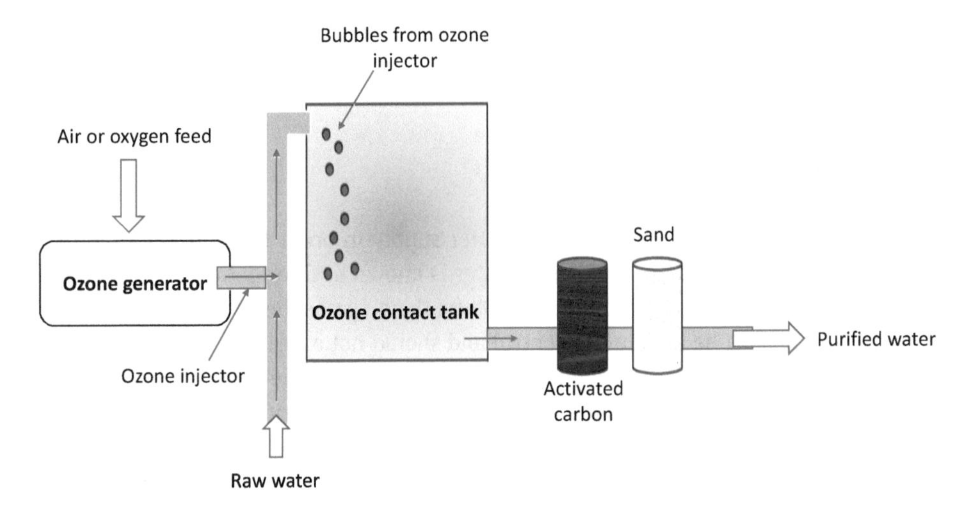

Figure 9.31: Water ozonation.

gas that completely removes flavor, color, and odor. This is due to the fact that ozone is unstable and decomposes into oxygen.

9.7.6 Photocatalytic Disinfection

Photocatalysis is nothing more than a photoreaction with a catalyst. It employs a light source to accelerate the disinfection process through the use of a catalyst. It is a complex oxidation technology that is both environmentally friendly and one of the most important green chemistry technologies today. This approach has a remarkable ability to kill a wide variety of dangerous microorganisms. Figure 9.32 depicts how photocatalysts employ solar energy to break down organic pollutants in water, thus cleansing wastewater efficiently and passively of dangerous substances. After cleaning the water,

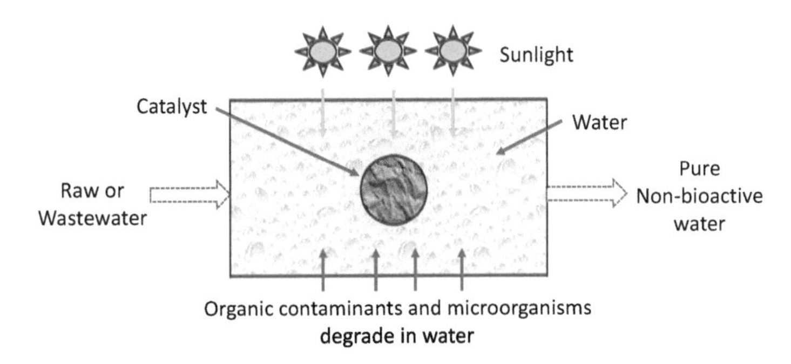

Figure 9.32: Photocatalytic disinfection.

the photocatalyst is completely recovered and can be used endlessly to produce additional clean water with no waste.

9.8 Fluoridation

The procedure of adding fluoride to the water supply in order to maintain a fluoride level of 0.7 ppm (milligrams) per liter of water is known as fluoridation. By demineralizing and remineralizing teeth, fluoride protects them against decay. Fluoride levels in drinking water must be kept under control and should not exceed a certain level, since too much fluoride might cause dental or skeletal fluorosis. The three chemicals most commonly used to fluoridate drinking water are sodium fluoride (NaF, Figure 9.33A), sodium fluorosilicate (Na_2SiF_6, Figure 9.33B), and fluorosilicic acid (H_2SiF_6, Figure 9.33C).

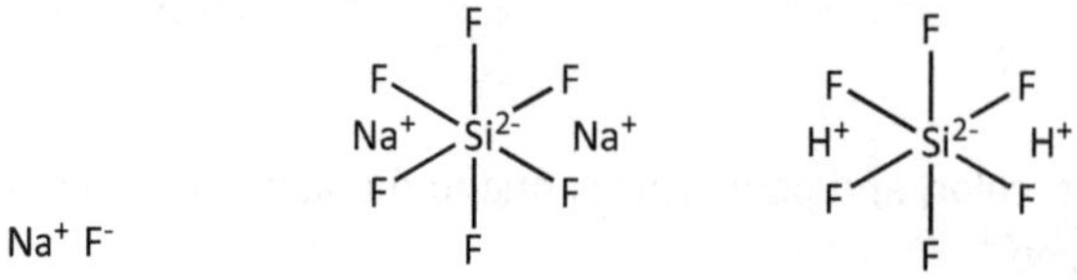

Figure 9.33: Chemicals used to fluoridate drinking water.

9.9 Questions

9.9.1 Provide some examples of organic coagulants.
9.9.2 What pH range is best for precipitation with aluminum salts?
9.9.3 What is the function of a coagulant aid?
9.9.4 Name three different types of sedimentation basins.
9.9.5 Define the term “water softening.”
9.9.6 Demonstrate how I-EX softening works.
9.9.7 Explain the process of zeolite regeneration.
9.9.8 Draw the chemical structure of fulvic acid, a chelating agent.
9.9.9 What is the effect of water in the upper zone of the Baylis curve?
9.9.10 Explain the water ozonation process in detail.

Chapter 10
Stormwater Management and Green Infrastructure

10.1 Introduction

Green infrastructure (GI) is a stormwater runoff control strategy that relies on natural processes to slow down, clean up, and occasionally reuse stormwater to keep it from overburdening sewage systems and hurting waterways. GI attempts to emulate the natural world by using plants, trees, and other features. It is considered as a preventative measure in the management of stormwater runoff contamination. The better the arrangement of GI, the better the management of stormwater (rainwater), the less runoff contamination, as shown in Figure 10.1.

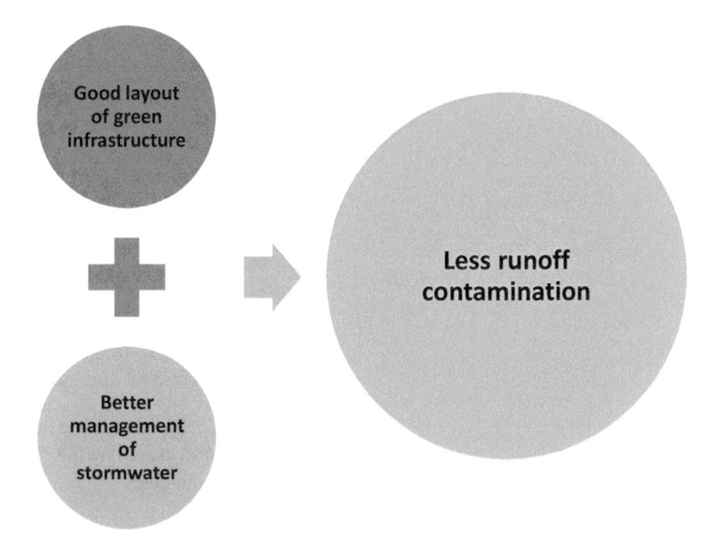

Figure 10.1: Stormwater management, green infrastructure, and runoff contamination relationship.

GI keeps waterways clean and healthy by capturing rain where it falls and allowing it to filter into the Earth, return to the atmosphere via evapotranspiration, or be reused for another purpose, such as landscaping. GI also improves water quality by lowering stormwater runoff into rivers and removing contaminants from that runoff.

Soil and plants help capture and remove pollutants from stormwater through adsorption, filtration, plant uptake, and organic matter decomposition. These mechanisms break down or trap many common pollutants in runoff, ranging from heavy metals to oil to microorganisms.

GI, in addition to affecting the quantity and quality of runoff that enters waterways, provides the following benefits, as summarized in Table 10.1.

https://doi.org/10.1515/9783111332468-010

Table 10.1: Green infrastructure benefits.

Lowering or decreasing flooding	Heavy rains are becoming more common and more intense. Climate change is exacerbating the problem by increasing the likelihood of flooding and sewer system overflows. Another growing issue is urban flooding induced by excessive rain on impermeable surfaces. Green infrastructure decreases flood hazards and increases community climate resilience by catching rain where it falls and diverting it away from sewers and rivers.
Enhancing water supply	More than half of the rain that falls in urban areas with impermeable surfaces ends up as runoff. Green infrastructure initiatives decrease runoff by absorbing stormwater and recharging groundwater sources or harvesting it for uses such as landscaping.
Reducing heat and smog	Vegetation is used in green infrastructure such as green roadways, green roofs, and other types to enhance air quality and reduce smog. In addition to offering shade, plants also aid in carbon sequestration by absorbing pollutants like carbon dioxide and reducing air temperatures through evapotranspiration and evaporation.
Strengthening public health	Green infrastructure has numerous positive effects on public health by lowering air pollution and temperatures, enhancing water quality, and enhancing the availability of open areas. In addition to encouraging physical exercise, greener spaces can enhance mental health while enhancing neighborhood livability.
Lowering expenses	Green infrastructure can be considerably less costly than more typical water management systems. It is less expensive to create a rain garden to address drainage difficulties than to build tunnels and install pipes. Green roofs can last twice as long compared to conventional roofs, and permeable pavement is an excellent long-term investment because it does not require maintenance.
Improving life quality	Green infrastructure creates jobs in numerous ways while also increasing the value of local real estate. Furthermore, pleasant, green environments encourage people to congregate outside.

10.2 Types of Green Infrastructure

GI can take numerous forms, from blossoming rooftop gardens to the absorbent pavement to tree-lined streets. Here are some typical examples:

10.2.1 Green Roofs

Green roofs are living vegetation settings that offer a lush haven for birds, butterflies, and humans. Green roofs reduce cooling and heating energy usage and expenditures by giving an extra layer of insulation to a home or structure. On hot summer days, a typical dark-colored roof transfers the heat it absorbs to the building below, whereas a green roof can stay cooler than the ambient air temperature around it. Figure 10.2 shows a basic design for these green roofs.

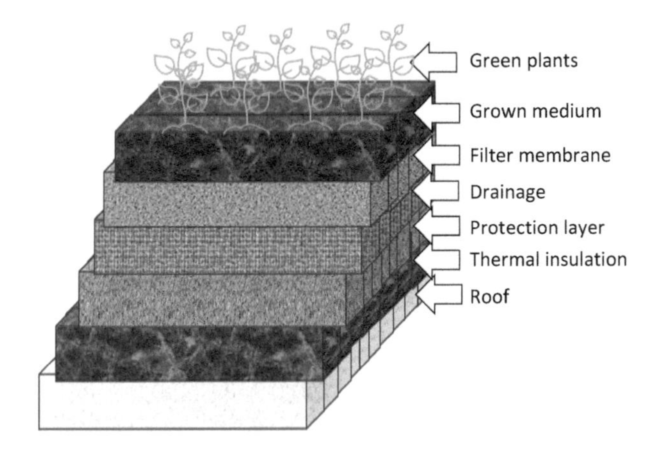

Figure 10.2: Green roof's basic design.

Green roofs also aid in the reduction of rain and carbon pollution. Furthermore, water is progressively released from a green roof, reducing the amount of runoff that enters a watershed all at once, and averting flooding and erosion. Additionally, green roof flora absorbs CO_2 and stores it in plants and soil.

10.2.2 Rain Gardens

Rain gardens are commonly composed of native plants, perennials, and grasses planted in a shallow basin and can be used in a variety of settings ranging from street medians to small yards as shown in Figure 10.3.

They are intended to catch and absorb roof, sidewalk, and street runoff. Rain gardens can also recharge underground aquifers, limit stormwater from reaching rivers, provide habitat for wildlife, and beautify a highway or yard in addition to allowing rainfall to evapotranspire or gently filter into the ground.

10.2.3 Permeable Pavement

Permeable pavement (also known as porous pavement) is a pavement system (Figure 10.4) which allows rainfall to seep through to underlying layers of pollutant-

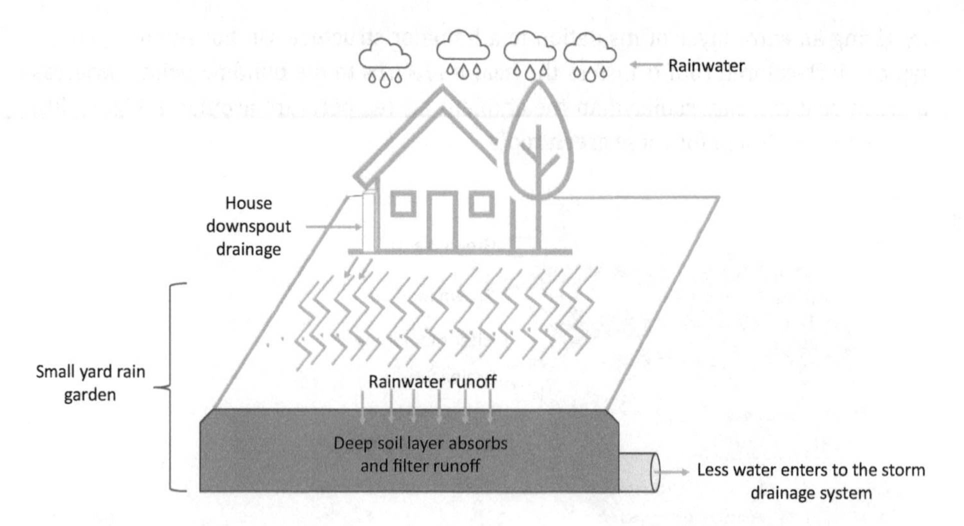

Figure 10.3: Small yard rain garden.

filtering soil before reaching groundwater aquifers. It is commonly used for sidewalks, parking lots, and driveways.

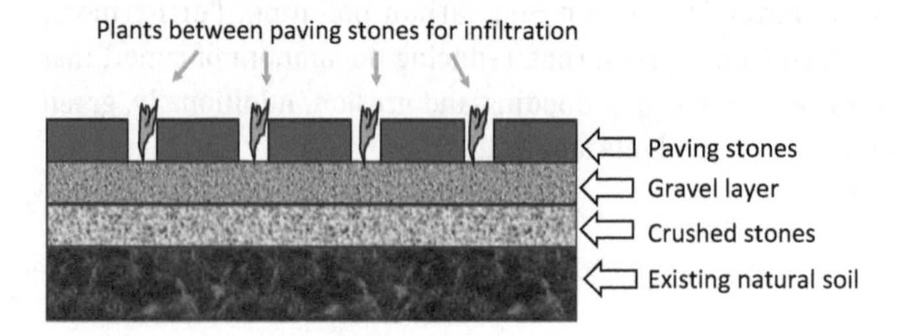

Figure 10.4: Permeable pavement layout.

10.3 Questions

10.3.1 Demonstrate the connection between stormwater management, GI, and runoff pollution.
10.3.2 What is stormwater runoff?
10.3.3 What causes stormwater runoff to be a problem?
10.3.4 What is GI?
10.3.5 Does GI cost more?

10.3.6 What are the primary goals of GI?
10.3.7 What are the three forms of GI?
10.3.8 What are the overall advantages of GI?
10.3.9 How does GI affect life quality?
10.3.10 Explain briefly how green roofs absorb rain and carbon emissions.

Chapter 11
Water as a Renewable Energy Source

11.1 Introduction

Renewable energy sources include any source that can deliver energy without depletion, as long as it is not used faster than it can renew. Wood, for example, is a renewable energy source if it is consumed at a rate equal to or lower than its rate of regeneration. Hydropower, often known as hydroelectric power, is a renewable energy source that creates electricity by altering the natural flow of a river or other body of water using a dam or a diversion construction. Water is considered renewable and produces the least solid waste during energy production when compared to other resources utilized to generate energy and power.

11.2 Endless Energy Sources

Air movement, sunlight, and water are the three primary sources of energy used in wind, solar, and hydroelectric power. These processes will continue continuously as long as there is life on Earth, meaning that energy can be taken from them. Because geothermal energy is generated by the heat of the planet's core, it is basically limitless. Unlimited energy sources, unlike renewable energy sources, never run out of energy.

11.3 Natural Resources

Minerals and other substances, classified as natural resources, are those that can be used for economic gain. Water, fertile soil, minerals, and forests are a few examples. Water and soil are two examples of natural resources that are necessary for life.

Natural resources come in both renewable and nonrenewable forms. The distinction between them is whether the resource can be replenished naturally while we are still alive or if it is permanently lost after being used up. As with trees in some areas or resources like oil and minerals that take hundreds, thousands, or even millions of years to refill, the fight with renewability is brought on by the realization that many of our resources are being used up too quickly to replenish themselves naturally.

https://doi.org/10.1515/9783111332468-011

11.4 Renewable Energy Resources

Crude oil will likely remain one of the most significant energy sources for decades, but the drive toward renewable energy has accelerated due to pollution from combustion engines, and depleting reserves. The rapid development of new renewable energy sources is a blessing. Table 11.1 lists a few of these renewable energy.

Table 11.1: A few renewable energy sources.

Biomass	Biomass fuels include agricultural products, wood, solid waste, landfill gas, biogas, and alcohol fuels such as ethanol or biodiesel.
Solar	Solar energy can now be converted and stored using today's batteries, resulting in an increase in solar plants and the usage of solar panels on anything, from street lamps to private residences.
Wind	Wind energy is being harvested all over the planet. Wind energy turbines require only an average wind speed of 15 km/h to generate energy. One commercial wind turbine can power up to 1,600 homes.
Geothermal	Geothermal energy generates energy and heat by utilizing the Earth's interior heat. This is especially important in volcanic areas or countries with extensive Sahara regions.
Hydropower or hydroelectric power	Hydroelectric energy, commonly referred to as hydroelectric power or hydroelectricity, is a form of energy that produces electricity by using the force of moving water, such as water flowing over a waterfall.

11.5 Hydroelectric or Hydropower Stations

Similar to factories, hydroelectric generating stations use the energy of moving or falling water to produce electricity or hydroelectric energy. A "drop," or passage of water from one elevation to another, generates speed and power. The natural drop of a river, like a waterfall or rapids, is frequently used by hydro stations as a source of water power. Other stations elevate the water level to provide the drop by building a dam across a river. A third kind, known as "run-of-river," uses only the river's natural flow and has no drop; hence having the least negative effects on the ecosystem.

The fundamentals of harnessing hydroelectric energy are largely unchanged despite several technological advances.

- Source: A dam, canal, or pump placed on, or close to, a moving body of water changes the direction of the water's flow.
- Harness: The water is redirected into a conduit that passes one or more turbines before coming out the other side and heading back in the original direction.

- Generate: The turbines are spun by the force of the flowing water and are connected to a generator, which harnesses the spinning power of the turbines to produce energy.
- Distribute: The electricity is provided to homes and businesses via the electrical grid, which uses only clean, renewable energy.

11.6 Types of Hydropower Energy Plants

To create energy, three basic types of hydroelectric power plants are used. Each type employs a distinct way of propelling a turbine and generating power for use by the grid. Table 11.2 summarizes these categories.

Table 11.2: Major types of hydroelectric power plants.

Impoundment	An impoundment system is the most typical type of hydroelectric energy. It uses a dam to fill a reservoir to store water and then releases the water to flow through a turbine to produce power.
Diversion	A diversion system diverts some water from its natural course and directs it into a canal or penstock where it flows through a turbine to produce energy. Diversion systems are frequently utilized in waterfalls and rivers that flow down from hills or mountains. However, they can also be used to harness the energy of ocean waves and tides.
Pumped storage	A natural battery is essentially what a hydropower system with pumped storage is. A pump that transfers water from a lower reservoir to an upper reservoir is powered by electricity produced elsewhere, as from a solar farm or a wind farm. The released water from the top reservoir is then sent via a turbine, which spins in order to produce energy.

11.7 Questions

11.7.1 Which types of energy can be used at any time?
11.7.2 What are some of the primary benefits and drawbacks of solar energy over fossil fuels?
11.7.3 What is the difference between conventional and nonconventional energy sources?
11.7.4 What exactly are hydroelectric or hydropower plants?
11.7.5 Give three examples of never-ending energy sources.
11.7.6 Which of the natural resources is required for life to exist?
11.7.7 What is the distinction between renewable and nonrenewable natural resources?
11.7.8 Give three examples of renewable energy sources.
11.7.9 Define geothermal energy in a few words.
11.7.10 What are the fundamentals of producing hydroelectric energy?

Essential Terms

Chelating agents: Organic compounds that may extract ions from aqueous solutions and convert them to soluble complexes.

Chemical contaminants: Substances that are toxic to aquatic plants and animals.

Chloramination: The process of disinfecting drinking water by adding chloramine to destroy viruses, bacteria, and other organisms that cause illness.

Coagulation: The use of coagulants and coagulant aids to precipitate colloidal turbidity particles.

Contaminant: Anything other than water molecules. Any physical, chemical, biological, or radiological substance or matter in water.

Corrosion: An electrochemical process that predominantly transforms metals and alloys into oxides, hydroxides, and aqueous salts.

Density: The weight of water per unit volume, which fluctuates with temperature.

Desalination: A common method to obtain fresh water by condensation of vapors produced by heating the seawater.

Disinfection: The eradication of water-borne microorganisms that enter the source water via sewage and runoff from the watershed.

Electrodialysis: A method of extracting ions from water by using current.

Euphotic zone: The warmest upper water layer that receives the most sunlight in the lake.

Filtration: Mechanical removal of turbidity or suspended particles from water by passing it through a porous material, such as a granular bed or a membrane.

Flocculation: The separation of a solution, commonly the removal of sediment from a fluid.

Flocculants: Substances that cause small particles in a solution to clump together, resulting in the development of a floc that floats to the surface or sinks to the bottom.

Fluoridation: The procedure of adding fluoride to the water supply in order to maintain a fluoride level of 0.7 ppm (milligrams) per liter of water.

Green roofs: Living vegetation settings that give a lush haven for birds, butterflies, and humans. Green roofs reduce cooling and heating energy usage and expenditures by giving an extra layer of insulation to a home or structure.

Green infrastructure: A stormwater runoff control strategy that relies on natural processes to slow down, clean up, and occasionally reuse stormwater to keep it from overburdening sewage systems and hurting waterways.

Hard water: Water that contains a high concentration of mineral ions.

Hardness: The amount of measured divalent metal cations.

https://doi.org/10.1515/9783111332468-012

Hydroelectric energy or hydroelectric power or hydroelectricity: A form of energy that produces electricity by using the force of moving water, such as water flowing over a waterfall.

Hydrophobic materials: Are "water-fearing." Do not combine or mix with water. They are insoluble in water.

Hydrophilic materials: Are "water-loving" and can be wetted by water. They are miscible with water.

Lentic water or lentic ecosystem: Aquatic system that moves so slowly that the water seems static or nearly so.

Limnetic zone: The lake's surface or open water area.

Littoral zone: The shoreline of a lake or pond.

Lotic water: The flowing waterways that have steady water movement with high oxygen concentration and clear water.

Maximum contaminant level: The highest level of a contaminant permitted in drinking water.

Maximum contaminant level goal: The highest contamination level in drinking water at which no known or predicted harmful effect on human health will occur, allowing for an adequate margin of safety.

Membrane filtration: The process of passing pretreated water under pressure through a membrane to remove particles of a certain size.

Microbial contaminants: Bacteria, viruses, and fungi.

Permeable pavement: A pavement system that allows rainfall to seep through to underlying layers of pollutant-filtering soil before reaching groundwater aquifers.

Radioactive waste: Any pollutant that emits more radiation than the environment normally does.

Rain gardens: Native plants and grasses planted in a shallow basin and used in a variety of settings ranging from street medians to small yards.

Reference dose: A safe dose based on research findings that have been extrapolated to humans from outcomes of animal tests.

Residual chlorine: The quantity of chlorine that remains in water after a specific period of time or contact time.

Reverse osmosis: A method of obtaining fresh water that involves filtering seawater under pressure via a semi-permeable membrane.

Safe drinking water act: Standards and regulations to provide access to safe, clean, and reliable drinking water, as well as adequate wastewater treatment.

Scale deposition: The buildup of various elements within undesirable locations.

Scum: A precipitate that is formed when hard water reacts with other substances in a solution, such as soap.

Sedimentation: The separation of flocs from flocculated water by gravity.

Sequestration: The ability to create a complex with metal ions that allows these ions to remain in solution in the presence of precipitation agents.

Sewage: The fraction of wastewater that contains feces or pee.

Spring: A location where underground water finds its way to the land surface and emerges.

Thermal stratification: A phenomenon that causes some lakes and ponds in temperate regions to divide into three distinct thermal layers or zones: the epilimnion, metalimnion, and hypolimnion.

Turbidity: A measure of how much transparency the water loses due to the presence of suspended particulates.

Water autoionization: The act of functioning in solution as both an acid and a base.

Water cycle: Is the constant movement of water both inside the earth and in the atmosphere.

Water hydrate: Any chemical that contains water in the form of H_2O molecules.

Water softening: A process of removing calcium, magnesium, and iron ions from water.

Water stabilization: A method of decreasing and controlling water contamination.

Well: A hole bored into the ground to gain access to water in an aquifer.

Zeolites: Microporous crystalline aluminosilicates that absorb water and other cations, filling the micropores.

Abbreviations

ARS	Acute radiation syndrome
$BaCl_2$	Barium chloride
$Ba(NO_3)_2$	Barium nitrate
$BaSO_4$	Barium sulfate
BGA	Blue-green algae
BOD	Biochemical oxygen demand
$CaCO_3$	Calcium carbonate
$CaNa_2EDTA$	Calcium disodium ethylenediaminetetraacetic acid
$Ca(OH)_2$	Calcium hydroxide (limewater)
CCL	Contaminant candidate list
CFU	Colony-forming unit
ClO_2	Chlorine dioxide
COD	Chemical oxygen demand
CNS	Central nervous system
DM	**Dem**ineralization
CRI	Cutaneous radiation injuries
D_2O	Deuterium oxide
DO	Dissolved oxygen
DNAPLs	Dense nonaqueous phase liquids
DPD	Diethyl-*p*-phenylene-diamine
DST	Defined substrate technology
EBT	Eriochrome Black T
EC	Electrical conductivity
ED	Electrodialysis
EDTA	Ethylenediaminetetraacetic acid
EHEC	Enterohemorrhagic *Escherichia coli*
EIEC	Enteroinvasive *Escherichia coli*
ETEC	Enterotoxigenic *Escherichia coli*
FTUs	Formazin turbidity units
GI	Green infrastructure
GMF	Granular medium filtration
HPC	Heterotrophic plate count
H_2SiF_6	Fluorosilicic acid
I-EX	**I**on **ex**change
LI	Langelier index
LNAPLs	Light nonaqueous phase liquids
LSI	Langelier saturation index
MED	Multiple-effect distillation
MF	Membrane filter
$MgCO_3$	Magnesium carbonate
$MgCl_2$	Magnesium chloride
$MgSO_4$	Magnesium sulfate
MCL	Maximum contaminant level
MCGL	Maximum contaminant level goal
MEE	Multiple effect evaporation
Mg/L	Milligrams per liter
MSF	Multistage flash

https://doi.org/10.1515/9783111332468-013

MTF	Multiple tube fermentation
MUG	4-Methylumbelliferyl-β-D-glucuronide
Na_2CO_3	Sodium carbonate (soda ash)
NaF	Sodium fluoride
NaOH	Sodium hydroxide
Na_2SiF_6	Sodium fluorosilicate
NFR	Nonfilterable residue
NTUs	Nephelometric turbidity units
ONPG	*o*-Nitrophenyl-β-D-galactopyranoside
PAH	Polycyclic aromatic hydrocarbon
PCB	Polychlorinated biphenyl
PCU	Platinum color units
DBPs	Potentially harmful disinfection byproducts
PPM	Parts per million
RfD	Reference dose
RNA	Ribonucleic acid
RO	Reverse osmosis
RSI	Ryznar Stability Index
SAC	Strong acid cation
SBA	Strong base anion
SDWA	Safe Drinking Water Act
SPC	Standard plate count
SPM	Suspended particulate matter
TDS	Total dissolved solids
TOC	Total organic carbon
TON	Threshold odor number
THMs	Trihalomethanes
TS	Total solid
TSM	Total suspended matter
TSS	Total suspended solids
TTN	Threshold taste number
UV	Ultraviolet
WHO	World Health Organization
VSS	Volatile suspended solids

Resources and Further Readings

Books

[1] Guidelines for Drinking-Water Quality, 4th Edition Incorporating the First Addendum. World Health Organization, 2017. ISBN 9789241549950.
[2] Darshan SS. Water Treatment Made Simple: For Operators. New Jersey: Wiley & Sons, Inc., 2006. ISBN: 9780471740025.
[3] Bitton G. Microbiology of Drinking Water: Production and Distribution. New Jersey: John Wiley & Sons, Inc., 2014. ISBN:9781118743928.
[4] Summers JK. Water Quality – Science, Assessments and Policy. London: IntechOpen Limited, 2020. ISBN 9781789855784.
[5] Meschke JS and Sobsey MD. Norwalk-like viruses: detection methodologies and environmental fate. In Encyclopedia of Environmental Microbiology, Bitton G (editor-in chief), New York: Wiley-Interscience, 2002, 2221–2235.
[6] Drinan JE and Spellman F. Water and Wastewater Treatment. A Guide for Non-engineering Professional, 2nd edition. Florida: CRC Press, 2012. eBook ISBN 9780429104879.
[7] Quevauviller P, Thomas O and Beken AVD. Wastewater Quality Monitoring and Treatment. New York: Wiley & Sons, Inc., 2006. ISBN: 9780471499299.
[8] Venkateswarlu KS. Water Chemistry, Industrial and Power Station Water Treatment. New Delhi: New Age International (P), Ltd., 1996. ISBN 9788122424997.
[9] Stoddard A, Harcum JB, Simpson JT, Pagenkopf JR and Bastian RK. Municipal Wastewater Treatment: Evaluating Improvements in National Water Quality. New York: Wiley & Sons, Inc., 2002. ISBN 9780471243601.
[10] Wetzel RG. Limnology: Lake and River Ecosystems, 3rd edition. San Diego: Academic Press, 2001. ISBN 9780127447605.

Journals

- Gleick PH and Palaniappan M. Peak water limits to freshwater withdrawal and use. Proceedings of the National Academy of Sciences of the United States of America, 2010, 107(25), 11155–11162. https://doi.org/10.1073/pnas.1004812107.
- Gliozzi A, Relini A and Chong PG. Structure and permeability properties of biomimetic membranes of bolaform archaeal tetraether lipids. Journal of Membrane Science, 2002, 206, 131–147. https://doi.org/10.1016/S0376-7388(01)00771-2.
- Greenlee LF, Lawler DF, Freeman BD, Marrot B and Moulin P. Reverse osmosis desalination. Water sources, technology, and today's challenges. Water Research, 2009, 43(9), 2317–2348. https://doi.org/10.1016/j.watres.2009.03.010.
- Khawaji AD, Kutubkhanah IK and Wie JM. Advances in seawater desalination technologies. Desalination, 2008, 221, 47–69. https://doi.org/10.1016/j.desal.2007.01.067.
- Chapman JM, Proulx CL, Veilleux MAN, Levert C and Bliss S, et al. Clear as mud: A meta-analysis on the effects of sedimentation on freshwater fish and the effectiveness of sediment-control measures. Water Research, 2014, 56, 190–202. https://doi.org/10.1016/j.watres.2014.02.047.
- Reid AJ, Carlson AK, Creed IF, Eliason EJ and Gell PA, et al. Emerging threats and persistent conservation challenges for freshwater biodiversity. Biological Reviews of the Cambridge Philosophical Society, 2018, 94(3), 849–873. https://doi.org/10.1111/brv.12480.

https://doi.org/10.1515/9783111332468-014

- Wyn-Jones AP and Sellwood J. Enteric viruses in the aquatic environment. Journal of Applied Microbiology, 2001, 91(6), 945–962. https://doi.org/10.1046/j.1365-2672.2001.01470.x.
- Rochman CM. Plastics and priority pollutants: multiple stressors in aquatic habitats. Environmental Science Technology, 2013, 47(6), 2439–2440. https://doi.org/10.1021/es400748b.
- Tian H, Alkhadra MA and Bazant MZ. Theory of shock electrodialysis I: Water dissociation and electroosmotic vortices. Journal of Colloid and Interface Science, 2021, 589, 605–615. https://doi.org/10.1016/j.jcis.2020.12.125.
- Wiegerinck HTM, Kersten R and Wood JA. Influence of charge regulation on the performance of shock electrodialysis. Industrial and Engineering Chemistry Research, 2023, 62(7), 3294–3306. https://doi.org/10.1021/acs.iecr.2c03874.
- Zhao Y, Zheng Y, Peng Y, Hea H and Sun Z. Characteristics of poly-silicate aluminum sulfate prepared by sol method and its application in Congo red dye wastewater treatment. RSC Advances, 2021, 11, 38208–38218. https://doi.org/10.1039/D1RA06343J.
- Ng M, Liana AE, Liu S, Lim M, Chow CWK, Wang D, Drikas M and Amal R. Preparation and characterisation of new-polyaluminum chloride-chitosan composite coagulant. Water Research, 2012, 46(15), 4614–4620. https://doi.org/10.1016/j.watres.2012.06.021.
- Gamage J and Zhang Z. Applications of photocatalytic disinfection. International Journal of Photoenergy, 2010, 2010, Article ID 764870. https://doi.org/10.1155/2010/764870.

Internet Resources

- Seawater desalination. Accessed on 01-05-2023. https://www.ctc-n.org/technologies/seawater-desalination.
- Water Pollution: Everything You Need to Know. Accessed on 08-05-2023. https://www.nrdc.org/stories/water-pollution-everything-you-need-know#prevent.
- Water pollutants. Accessed on 05-05-2023. https://www.pca.state.mn.us/air-water-land-climate/water-pollutants.
- Water Microbiology. Accessed on 09-05-2023. https://onlinelibrary.wiley.com/ doi/10.1002/9781118743942.ch10.
- Water Quality Parameters. Accessed on 30-05-2023. https://www.intechopen.com/chapters/69568.
- Water Chemistry. Accessed on 30-05-2023. https://www.intechopen.com/books/8178.
- Testbook. Accessed on 30-05-2023. https://testbook.com/question-answer/activated-carbon-is-used-for--63d97beb31adf2a06f9a50f3.
- Types of Drinking Water Contaminants. Accessed on 12-06-2023. https://www.epa.gov/ccl/types-drinking-water-contaminants.
- Contaminant Candidate List (CCL) and Regulatory Determination. Accessed on 12-06-2023 https://www.epa.gov/ccl/contaminant-candidate-list-1-ccl-1.
- The Sources of Chemical Contaminants in Food and Their Health Implications. Accessed on 13-06-2023. https://www.frontiersin.org/articles/10.3389/fphar.2017.00830/full.
- Unique properties of water. Accessed on 14-06-2023. https://www.khanacademy.org/science/ap-biology/chemistry-of-life/structure-of-water-and-hydrogen-bonding/a/hs-water-and-life-review.
- Properties of Water. Accessed on 14-06-2023. https://bio.libretexts.org/Bookshelves/Introductoryand_General_Biology/Map%3A_Raven_Biology_12th_Edition/02%3A_The_Nature_of_Molecules_and_the_Properties_of_Water/2.05%3A_Properties_of_Water.
- Hard Water. Accessed on 15-06-2023. https://chem.libretexts.org/Bookshelves/Inorganic_Chemistry/Supplemental_Modules_and_Websites_(Inorganic_Chemistry)/Descriptive_Chemistry/Main_Group_Reactions/Hard_Water.

- Heavy water. Accessed on 16-06-2023. https://en.wikipedia.org/wiki/Heavy_water.
- The Natural Sources of Water. Accessed on 16-06-2023. https://www.aquafil.com.au/natural-sources-water.
- Rainwater. Accessed on 07-07-2023. https://www.sciencedirect.com/topics/engineering/rainwater.
- Rainwater harvesting. Accessed on 07-07-2023. https://sustainabletechnologies.ca/home/urban-runoff-green-infrastructure/low-impact-development/rainwater-harvesting.
- Aquatic Science Lotic Ecosystems. Accessed on 08-07-2023. https://slideplayer.com/slide/8256457.
- Threats on Aquatic Ecosystem- Mitigation and Conservation Strategies. Accessed on 08-07-2023. https://www.heraldopenaccess.us/openaccess/threats-on-aquatic-ecosystem-mitigation-and-conservation-strategies.
- Groundwater. Accessed on 08-07-2023. https://www.epa.gov/sites/default/files/documents/groundwa ter.pdf.
- Causes and Effects of Agricultural Pollution. Accessed on 10-07-2023. https://www.earthreminder.com/causes-and-effects-of-agricultural-pollution.
- Oil spills. Accessed on 10-07-2023. https://www.noaa.gov/education/resource-collections/ocean-coasts /oil-spills.
- Groundwater Contamination. Accessed on 11-07-2023. https://groundwater.org/threats/contamination/.
- Water Microbiology. Bacterial Pathogens and Water. Accessed on 12-07-2023. https://www.ncbi.nlm.nih.gov/pmc/articles/PMC2996186.
- Water Analysis. Accessed on 13-07-2023. https://www.sciencedirect.com/topics/agricultural-and-biological-sciences/water-analysis.
- Experiment-8 to determine odor and taste in water sample. Accessed on 13-07-2023. https://nitsri.ac.in/Department/Civil%20Engineering/CIV-701_P_Water_Quality_Lab_Exp_8.pdf.
- Get Informed Color, Taste, and Odor (Smell). Accessed on 13-07-2023. https://www.knowyourh2o.com/indoor-6/color-taste-odor#:~:text=Odor%20(Smell) %20%2D%20Phenolic%2F, Hydrogen %20Sulfide%2C%20microbiological %20 contaminants%3B%20Perfume.
- How to Measure Turbidity of Water. Accessed on 13-07-2023. https://waterfilterguru.com/how-to-measure-turbidity-of-water.
- What Is Alkalinity? Accessed on 15-07-2023. https://blog.orendatech.com/what-is-alkalinity.
- Water Hardness. Accessed on 17-07-2023. https://www.sciencedirect.com/topics/biochemistry-genetics-and-molecular-biology/water-hardness.
- Temporary and Permanent Hardness of Water. Accessed on 17-07-2023. https://watertreatmentbasics.com/temporary-and-permanent-hardness-of-water.
- What Is Residual Chlorine? Accessed on 19-07-2023. https://apureinstrument.com/blogs/what-is-residual-chlorine.
- Test for Residual Chlorine. Accessed on 19-07-2023. https://www.thewatertreatments.com/disinfection/test-residual-free-chlorine.
- The Winkler Method – Measuring Dissolved Oxygen. Accessed on 21-07-2023. https://serc.carleton.edu/microbelife/research_methods/environ_sampling/oxygen.html.
- Green Infrastructure: How to Manage Water in a Sustainable Way. Accessed on 21-07-2023. https://www.nrdc.org/stories/green-infrastructure-how-manage-water-sustainable-way#whatis.
- What's in Your Water: Total Dissolved Solids (TDS) in Drinking Water. Accessed on 21-07-2023. https://quenchwater.com/blog/tds-in-drinking-water.
- Total Suspended Solids. Accessed on 22-07-2023. https://www.rocker.com.tw/en/application/sus pended_solid_test.
- How to Test Your Well Water for Coliform Bacteria – The Invisible Danger. Accessed on 23-07-2023. https://www.health-metric.com/blogs/water-quality-blog/how-to-test-your-well-water-for-coliform-bacteria-the-invisible-danger#:~:text=Testing%20your%20well%20water%20for%20coliform%20bacter ia&text=The%20at%2Dhome%20test%20uses,is%20positive%20for%20coliform%20bacteria.

- Membrane Filtration Method. Accessed on 23-07-2023. https://biologyreader.com/membrane-filtration-method.html#:~:text=Purpose%3A%20The%20main%20objective%20of,of%20colonies%20through%20colony%20counter.
- Presence Absence (P-A). Accessed on 23-07-2023. https://www.uwyo.edu/molb2021/virtual-edge/lab22/exp_22a_pa.html.
- Heterotrophic Plate Count. Accessed on 23-07-2023. https://water.mecc.edu/courses/ENV295Micro/lesson9.htm.
- Three Main Types of Water Quality Parameters Explained. 20 Parameters. Accessed on 24-07-2023. https://sensorex.com/three-main-types-of-water-quality-parameters-explained/#:~:text=There%20are%20three%20water%20quality,chemical%20parameters%2C%20and%20biological%.
- Water Quality Parameters. Accessed on 24-07-2023. https://www.ysi.com/parameters.
- Color measurement. Accessed on 25-07-2023. https://assets.thermofisher.com/TFS-Assets/LPD/Application-Notes/an_034_tip_color_measurement_1120.pdf.
- Hydrogen Peroxide – Wastewater. Accessed on 25-07-2023. https://www.usptechnologies.com/hydrogen-peroxide/.
- Inorganic Coagulants for Water Treatment. Accessed on 25-07-2023. https://www.kemira.com/products/inorganic-coagulants-for-water-treatment/?utm_source=google&utm_medium=ad&utm_campaign=iw-chemicals&utm_content=coagulants-offering&gclid=CjwKCAjwlJimBhAsEiwA1hrp5nW7_NYPjXAIhBIkBGmQEBL7YDPIuVqIDhnGwEBJKkISTWRmZllV5hoCiSMQAvD_BwE.
- What Is Sedimentation in Water Treatment? Accessed on 31-07-2023. https://aosts.com/what-is-sedimentation-in-water-treatment-types-settling-tanks/
- Solids Contact Clarifiers. Accessed on 02-08-2023. https://www.monroeenvironmental.com/water-and-wastewater-treatment/circular-clarifiers-and-thickeners/solids-contact-clarifiers/
- Water Softening. Accessed on 02-08-2023. https://www.britannica.com/technology/water-softening.
- Manufacturers Require Additional Water Balance Index. Accessed on 02-08-2023. https://www.poolspanews.com/how-to/maintenance/manufacturers-require-additional-water-balance-index_o.
- Scaling. Accessed on 03-08-2023. https://www.cmu.edu/cee/research/cooling/scaling.html.
- Corrosion. Accessed on 03-08-2023. https://saylordotorg.github.io/text_general-chemistry-principles-patterns-and-applications-v1.0/s23-06-corrosion.html.
- Demineralization Plants (DM Plants). Accessed on 03-08-2023. https://soniwatercare.com/demineralization-plants/.
- Fundamentals of granular media filtration. Accessed on 07-08-2023. https://ca-nv-awwa.org/canv/downloads/sc16/Session9/pdfs/FundamentalsGranularMediaFiltration.pdf.
- Disinfection Methods. Accessed on 07-08-2023. https://www.intechopen.com/chapters/63788.
- Water Treatment. Accessed on 08-08-2023. https://www.frankgulab.com/water-treatment.
- Hydroelectric power: how it works. Accessed on 08-08-2023. https://www.opg.com/powering-ontario/our-generation/hydro/how-it-works/.
- What Is Hydroelectricity? Accessed on 08-08-2023. https://www.greenmountainenergy.com/why-renewable-energy/renewable-energy-101/hydro.
- How Is Water a Renewable Resource? Accessed on 09-08-2023. https://sciencing.com/about-5251373-water-renewable-resource-.html.
- Green Infrastructure Plan. Accessed on 10-08-2023. https://www.dcwater.com/green-infrastructure.
- Layers of a green roof. Accessed on 10-08-2023. https://blog.denbow.com/greenroof-layers.
- 5 Steps for Creating a Rain Garden. Accessed on 10-08-2023. https://www.hillsboroughcounty.org/en/newsroom/2018/04/10/a-rain-garden-is-an-attractive-way-to-improve-water-quality.
- Permeable Paving. Accessed on 10-08-2023. https://water.phila.gov/gsi/tools/permeable-paving/.

Index

https://doi.org/10.1515/9783111332468-015